Geometric Cosmos
Geometric Universe

Geometric Unification of Coupling Constants and Masses
Fermion Masses as Powers of e, π and 2
Separation of Fermion Masses into Sequences
Coupling Constants as Powers of e and 2
Internal Particle Universe Shells
Gavitation G Coupling Constant Value
Quantization of Universe Masses

Stephen Blaha Ph. D.
Blaha Research

Pingree-Hill Publishing
MMXXIV

ISBN: 979-8-9894084-6-7

Rev. 00/00/01 July 14, 2024

To Margaret

Some Other Books by Stephen Blaha

SuperCivilizations: Civilizations as Superorganisms (McMann-Fisher Publishing, Auburn, NH, 2010)

All the Universe! Faster Than Light Tachyon Quark Starships & Particle Accelerators with the LHC as a Prototype Starship Drive Scientific Edition (Pingree-Hill Publishing, Auburn, NH, 2011).

Unification of God Theory and Unified SuperStandard Model THIRD EDITION (Pingree Hill Publishing, Auburn, NH, 2018).

The Exact QED Calculation of the Fine Structure Constant Implies ALL 4D Universes have the Same Physics/Life Prospects (Pingree Hill Publishing, Auburn, NH, 2019).

Passing Through Nature to Eternity ProtoCosmos, HyperCosmos, Unified SuperStandard Theory (Pingree Hill Publishing, Auburn, NH, 2022).

HyperCosmos Fractionation and Fundamental Reference Frame Based Unification: Particle Inner Space Basis of Parton and Dual Resonance Models (Pingree Hill Publishing, Auburn, NH, 2022).

The Cosmic Panorama: ProtoCosmos, HyperCosmos, Unified SuperStandard Theory (UST) Derivation (Pingree Hill Publishing, Auburn, NH, 2022).
Ultimate Origin: ProtoCosmos and HyperCosmos (Pingree Hill Publishing, Auburn, NH, 2022).

God and and Cosmos Theory (Pingree Hill Publishing, Auburn, NH, 2023).

Newton's Apple is Now The Fermion (Pingree Hill Publishing, Auburn, NH, 2023).

Cosmos Theory: The Sub-Particle Gambol Model (Pingree Hill Publishing, Auburn, NH, 2023).

Cosmos-Universe-Particle-Gambol Theory (Pingree Hill Publishing, Auburn, NH, 2024).

Fractal Cosmos Curve: Tensor-based Cosmos Theory (Pingree Hill Publishing, Auburn, NH, 2024).

*The Eternal Form of Cosmos Theory Third Edition (*Pingree Hill Publishing, Auburn, NH, 2024).

*Fundamental Constants of Cosmos Theory and The Standard Model (*Pingree Hill Publishing, Auburn, NH, 2024).

Available on Amazon.com, bn.com Amazon.co.uk and other international web sites as well as at better bookstores.

CONTENTS

FIGURES and TABLES

Introduction

Cosmos Theory has a sequence of spaces. Ten spaces are Physical in the sense of supporting universes. Previously, we found sequences of Coupling Constants, of up-type fermion masses, and of down-type fermion masses. We now extend Cosmos Theory to sequences of universe masses at the time of universe creation and to the Gravitation constant G.

This book covers many topics. It considers these topics from a deeper perspective in Mathematics and Geometry. The book suggests fermions have inner universes that explain their masses.

The sequence of Cosmos spaces is the result of the number of totally antisymmetric tensors of dimension r and the number of creation and annihilation operators in a Fermion wave function – Mathematics!

Coupling constants form a sequence in powers of two – they are a product of the number of fermion spins in a symmetry group representation – Mathematics and Geometry – based on a product of degrees of freedom and multi-dimensional geometry volume and surface values.

The two sequences of fermion masses that the author found are the result of an internal universe shell for each fundamental fermion. Up-type fermions have 5, 4, 3, and 2 dimension internal universe shells. Down-type fermions have 3 dimension internal universe shells. The shells generate powers of $\pi = 3.14159$ and $e = 2.718$ Geometrically.

They implement the mass multipliers:

Up type fermion multiplier: 32π
Down type fermion multiplier: 32

The 32 multiplier appearing in both sequences is the number of Normal (non-Dark) fermions in a fermion layer.

The appearances of e and π are explained as due to volume and surface factors in the fermion mass based on a universe consistency condition applied for fermion stability – Geometry!

The form of the e and π mass entries is the reason for viewing fermions as having internal universe shells based on their variations. These π and e variations, based on internal universe shell dimension, are "explained" by volume and surface area factors – Geometry!

The sequence of universe total mass-energies E is based on a universe creation consistency condition – a consistency condition for the creation of a universe where the pressure in the universe equals the Casimir force generated by the enclosing universe. The total mass-energy has a joint sequence in powers of 2, of the enclosing universe's dimension, and of the universe's dimension r. The basis of its form is the form of surface area and volume in various dimensions, and of Cosmos Theory dimension array sizes (numbers of fermions).

These results suggest that a universe is a form of particle.

The gravitation coupling constant G is associated with the SU(256) symmetry group. A mass factor is required on dimensional grounds in gravitational dynamic equations. This factor causes gravitation G to be weak. We now have a viable estimate of the value of G and its origin in a coupling constant formulation.

The seven sequences that appear: Cosmos spaces, Coupling Constants, Gravitation coupling constant, Up-Type fermions, Down-Type fermions, Vector Boson W and Z masses, and universe masses are all based on Mathematics and Geometry. See chapter 24.

We conclude the total Lagrangian for our universe is significantly fixed:

1. Cosmos Theory gives spaces and dimension arrays, coupling constants and masses.

2. The particles Lagrangian is almost totally defined by using coupling constants and masses (and scalar and vector bosons).

A fait Accompli!

The book's list of evidence for Cosmos Theory is further augmented by the calculation of the Weinberg angle and the vector boson Z and W masses. The book also makes projections of fermion masses to as yet unknown higher generations.

Origin of the Cosmos Theory Spaces

1. The Origin and The End of the Cosmos

There have been numerous speculations about the Beginning and End of all things. These speculations assume that a space and time exists that can be used to discuss beginnings and endings. We chose to view this assumption with suspicion. Space and time are constructs based on the changes in substances. Where there is no substance they have no meaning.

We assume substances exist only within universes. Space and time within a universe have a beginning and an end – the lifetime and extent of the universe. It is only possible to discuss a before-time or a region beyond the spatial extent of a universe if the universe exists within a larger universe. Thus we have the possibility of before and after, and in and beyond for a universe—only for "contained" universes.

Two possibilities present themselves: there is an infinite chain of nested universes as some propose OR there are one or more primeval Parent universes within which all "child" universes reside. We chose the latter possibility in our definition of the Cosmos.

In Cosmos Theory we may assume that one ultimate Parent universe exists that we chose to be the seed universe. It has eighteen space-time dimensions. We may then assume "child" universes may exist within it including our own universe. Grand children universes and so on may exist too.

The parent universe specifies the time and spatial extents for its child universes. There is no question of "before" or "after" for the parent universe. All child universes may have their times and extents set relative to their time and extent within the parent universe.

These observations are based on:

1. There can be no dynamics if there is no time or space.
2. Time and distances only exist for substances (including vacuums containing gravitation.)
3. Measures of absolute motion and location are meaningless outside of universes.

1.1 Cosmos Structure

It is possible to consider the Cosmos as composed of universes with random contents and dynamics. We chose to impose order on the set of universes by assuming universes comply with a consistent, related structure of Cosmos Theory. We determined this Cosmos structure from a consideration of the nature of dimensions and coordinates and related quantities such as integrals over the space and time of substances (entities).

Therefore we view the definition of Cosmos Theory as initially beginning with a specification of the general structure of the set of spaces composing the Cosmos. Then universes may be defined or created through dynamical assumptions or mechanisms. We have chosen an ultimate, single Parent universe – an eighteen space-time dimension universe from which all other universes are descended although other possibilities present themselves.

There is no time before this Parent universe(s) and no time after this universe(s) should it end. Cosmos structure exists for all time and space within the Parent universe. It is eternal in Cosmos Theory. Therefore, previous attempts to generate a Cosmos structure such as ProtoCosmos Theory dynamically are beside the point. There is no dynamics before time and space exists in a Parent universe.

1.2 Lifetime of Cosmos Structure and Parent Universe(s)

Our discussions of Cosmos features in the text are necessarily sequential. The structure and content of Cosmos Theory is "instantaneous." The structure of Cosmos Theory spaces exists at the same instant as the Parent universe(s). (There may be more than one Parent universe. Each such universe would be at the top of a hierarchy of possible subuniverses.) The lifetime of a Parent universe must be viewed as zero since there is no external clock to specify a time. The lifetime of a Parent universe is simultaneous with the lifetime of the (eternal) Cosmos structure. The lifetime of a subuniverse depends on universe dynamics.

Within the Parent universe evolution is possible in a dynamics based on the internal Parent time. Thus subuniverses may be generated dynamically through inter-universe dynamics.

1.3 "Eternal" Nature of Cosmos Structure

We have arrived at an open door: the Nonexistence (Nothingness) of Cosmos Theory's origin and structure. It is eternal since there is no time to measure it. The Cosmos structure is not a substance. It only exists to specify the form of Cosmos spaces.

Subuniverses (child universes) have substance. They have attributes such as mass, energy, internal coordinate systems, and internal structures.

Substances exist within universes. They generally have finite extent and lifetime. They may be called "mortal."

Cosmos Structure is eternal but not substantial. We have shown that its nature is based directly on Mathematics.

1.4 Is What Creates the Cosmos Universes?

There appear to be several points in Cosmos Theory where a creator[1] may act. A creator can define the form of the Cosmos Theory set of spaces. This possibility is mitigated somewhat by the mathematical form from which Cosmos Theory is derived. (See the following chapters.) The form, which we derive, may be the only possible form that is without inconsistencies at some level. It also has the virtue of being directly derived from fundamental Mathematical concepts.

---

[1] We view the creator as an instrumentality that initiates a process.

A creator might be the creator of Parent universes, from which all subuniverses flow. There is the question of whether a creator is required since the structure of Cosmos Theory is defined simultaneously with the Parent universe(s) and would end at the same instant as the Parent universe(s) ends.[2]

Within a subuniverse a creator might act. There is no Physical requirement for such interactions.

1.5 Introduction of Quantum Theory

The correct view of Quantum Theory is to regard it as the result of Quantum Field Theory as this author has stated a number of times previously. The Physics of Quantum Mechanics is derivable from Quantum Field Theory.[3]

Therefore the definition of Lagrangian Quantum Field Theories of Physics leads inexorably to the experimentally known Quantum Phenomena. For example Quantum Electrodynamics is the heart and soul of photons, atoms, and their electromagnetic character and interactions.

On this basis we can view Cosmos Theory as implementing Quantum Theory through its use of Lagrangian Quantum Field Theory. Our work uses conventional quantum field theory and our generalization to PseudoQuantum Theory, which has conventional quantum field theory as a subset.

1.6 Fundamental Prerequisites for Cosmos Theory

The views expressed in this chapter are based on:

1. There can be no time or space or dynamics outside of universes.
2. The structure of Cosmos Theory is not substantial.
3. Cosmos Theory contains an overall design that governs the form of spaces and universes.
4. Universes cannot interact with each other unless they are both within a larger universe.
5. The form of Cosmos Theory is solely based on Mathematical considerations that were presented in earlier books by the author, and will be discussed in more detail later in this book.

---

[2] If there are more than one Parent universe, then all Parent universes have the same Cosmos structure by assumption. Parent universes cannot interact because there is no intervening space and time between them. Parent universes are totally independent of each other. Their sets of subuniverses are also independent of each other for the same reason. Multiple Parents constitute totally separate Cosmos's. We will assume only one Parent universe. If other Cosmos Parents exist would be very similar, if not identical, in structure and general features.

[3] Some theorists like to entertain the concept of an independent Quantum Mechanics based on a set of postulates. This approach is mathematically acceptable. But it is not Physics. It is Applied Mathematics. A basis in Quantum Field Theory is necessary for a consistent theory of Physics.

2. Dimensions, Coordinates, and Antisymmetric Tensors

This chapter establishes the Mathematical framework for the derivation of the structure of Cosmos Theory spaces embodied in space dimension arrays. It begins by considering dimensions and sets of coordinates. The considerations of multiple integrations over the spatial extent of objects lead to n-forms, object orientations, and the asymmetric form of wedge products and antisymmetric tensors.

One goal is to develop the basis of γ-matrices (Dirac γ-matrices and γ-matrices in higher dimensions) and the corresponding basis for Cosmos Theory dimension arrays.

The main goal is to derive the structure of the set of Cosmos spaces from fundamental mathematical considerations. We have found the spectrum of spaces' dimension array sizes to be a sequence of powers of 2.

2.1 Dimension

The concept of dimension requires clarification. We shall begin the discussion by regarding dimension as an ordered set of integers extending from $-\infty$ to $+\infty$. Each integer in the set is a dimension and may be associated with a set of coordinates.

2.2 Coordinate Systems

The nth dimension corresponds to n coordinates that vary over a range of values. These coordinates may be labeled from 1 to n. For each n there is a coordinate system.

2.3 Integration Elements

In any coordinate system one may consider integrations (summations) over the region occupied by an object. Integrations may be defined using volume and area elements of the form

$$dx_1 dx_2 \ldots dx_k \quad (2.1)$$

for some value $k \leq n$. The factors in each element, if treated as independent integration elements, are "classical." They commute.

2.4 Orientation of Objects

One may introduce an orientation for integration over an object. The object in Fig. 2.1 is shown as oriented. There are rules for orientation specification. One rule states the orientation is specified by "walking around the edge of the object keeping the left hand inside the object." Orientation is usually counterclockwise. Another rule for specifying orientation uses a normal to the object pointing "up" from the object. The orientation of an object enables integrations to embody orientation by the ordering of the factors of the integration element.

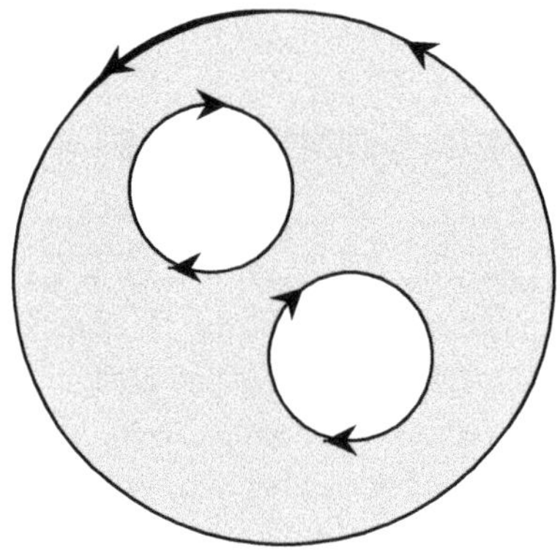

Figure 2.1. An Oriented object.

2.5 Antisymmetry of Oriented Elements

For example, if the object is oriented then

$$dxdy = -dydx$$

Oriented differential integration elements have an antisymmetry. For higher dimension differential elements we may obtain totally antisymmetric integration elements.

2.6 n-forms

The totally antisymmetric integration elements lead to totally antisymmetric n-forms. The set of all differential forms for dimension n consists of n + 1 forms that are named 0-form, 1-form, 2-form, … , n-form. A k-form has k factors corresponding to the differential form $dx_1dx_2 \ldots dx_k$. The 0-form corresponds to an integration with no differential form factor, namely, non-integration.

The k-forms are independent and totally antisymmetric in their factors.

2.7 Structure of n-Forms

If one considers the set of k-forms for a space of dimension n then one sees that may be enumerated due to their antisymmetry.[4] For example if n = 4 we find the following list of antisymmetric k-forms (listed by index):

0-form:	1 item	-
1-form	4 items	1 2 3 4
2-form	6 items	12 13 14 23 24 34
3-form	4 items	123 124 134 234
4-form	1 item	1234

The total number of items is $16 = 2^4$ where the power of "4" is the dimension. Note the antisymmetry constrains the set of items for each k-form.

---

[44] One may also do enumerations based on other considerations such as, perhaps, totally symmetric tensors. They total to $6^{r/2}$ independent tensors for dimension r. These enumerations are not of interest here.

Examining the number of items for each k-form above we see they are binomial coefficients:

$$\begin{bmatrix} r \\ k \end{bmatrix}$$

where r is the dimension of the space. We see r = 4 in the above example.

In general the sum of the (r + 1) k-forms of dimension r

0-form, 1-forms, …. r-form

is[5]

$$2^r \quad (2.2)$$

2.7.1 Internal Structure of n-Form Distribution

The above example, and the general case, exhibits a regularity. The items of an n-form are a quadruple (suitably arranged) of the (n – 2)-form. For the above n = 4 case we find

n=4	1	4	6	4	1	

is the ordered sum of four n-2 forms

	1	2	1			
		2	4	2		TIMES 2
			1	2	1	TIMES 1
TOTAL	1	4	6	4	1	

The summation requires offsets of the n-2-form items. We can specify the rule as

$$T_n = T_{n-2} + 2\ T_{n-2}(1) + T_{n-2}(2) \quad (2.3)$$

where the second and third summand is offset by the number within the parentheses. The total number of n-form elements is

$$2^n \quad (2.3a)$$

Another summation that generates an n-form combines (n – 1)-forms:

$$T_n = T_{n-1} + T_{n-1}(1) \quad (2.4)$$

---

[5] In an r dimension space the number of integrand factors for objects: lines, areas, and volume elements must be less then r + 1.

The quadrupling effect carries over to γ-matrices and dimension arrays seen later.

2.8 Wedge Product Representation of n-forms

A k-form may be represented using wedge operators, ∧. Wedge operator expressions are totally antisymmetric. A wedge representation of a k-form is

$$dx_1dx_2 \ldots dx_k \equiv dx_1\wedge dx_2\wedge \ldots \wedge dx_k \tag{2.5}$$

Wedge expressions satisfy the rules

1. Total antisymmetry. For example
$$dx \wedge dy = - dy \wedge dx$$
2. Any wedge product with a pair (multiple) of identical factors is zero. For example
$$dx \wedge dx = 0$$

2.9 Totally Antisymmetric Tensors

The above discussion of n-forms has an equivalent when one considers tensors. The tensors of an r dimension space with indices xyz… have the form:

$$T_{xyz \ldots}$$

Totally antisymmetric tensors have significant Physical applications such as the determination of the form of Dirac γ-matrices in 4 dimensions. Tensor indices range from a zero index tensor through an r index tensor. Thus there are r + 1 types of antisymmetric tensors in r dimensions.

2.10 Number of Independent Antisymmetric Tensors in Even Number Dimensions

The number of independent,[6] totally antisymmetric tensors with k indices in an even r dimension space is the same as the number of items for each k-form seen above. The "number of items" numbers are binomial coefficients:

$$\begin{bmatrix} r \\ k \end{bmatrix}$$

There are r + 1 types of tensors: those tensors with 0 indices, tensors with one index, …, tensors with r indices. The total number of independent antisymmetric tensors obtained by summing the r + 1 binomial coefficients is

$$2^r \tag{2.6}$$

where r is the dimension of the space.

---

[6] In an odd dimension space tensors are interrelated and as a result the number of independent antisymmetric tensors is 2^{r-1}.

3. Degrees of Freedom of γ-Matrices and Dimension Arrays

The discussion of tensor forms in chapter 2 leads to an understanding of the form of γ-matrices in four and other dimensions. First we note the number of tensors is the same as the number of independent γ-matrices *and* related matrices in dimension r. We only consider the case of even number dimensions.[7]

The number of independent, antisymmetric tensors in even dimension r is given by eq. 2.6. This number is the number of independent components for γ-matrices and related matrices in dimension r. Thus the γ-matrices have 2^r components. Since they are square matrices they have $2^{r/2}$ rows and $2^{r/2}$ columns. For example, Dirac matrices are 4 by 4 matrices in our four dimension space.

3.1 Fermion Spin

We now relate the structure of γ-matrices to the spin of fermions in even number dimension r. The key fact is the number of spins for a dimension is half of the column length of a γ-matrix:

$$\text{Number of spins} = 2^{r/2-1} \tag{3.1}$$

The lowest spin s of a fermion in dimension r is determined by

$$2s + 1 = \text{Number of spins} = 2^{r/2-1} \tag{3.2}$$

Therefore the lowest fermion spin in the even dimension r is[8]

$$s = 2^{r/2-2} - ½ \tag{3.3}$$

If r = 4, then s = ½ .

3.2 Fermion Quantum Fields

Having determined the lowest spin of fermions in dimension r we may now turn to the form of fermion quantum fields in that dimension. A Fourier representation

$$\psi(x) = \Sigma_{\alpha,s}[b(\alpha, s)u_{\alpha s}f_\alpha(x) + d^\dagger(\alpha, s)v_{\alpha s}f_\alpha{}^*(x)] \tag{3.4}$$

---

[7] As we pointed out previously the odd number dimension cases have the same results as the next lowest even number dimension case. When restricting ourselves to Cosmos Theory, we avoid duplication by considering even number dimensions only.

[8] Those concerned with perturbation theory divergences in higher dimensions are reminded that the Two Tier Theory of the author removes all such divergences for any dimension.

where α represents the Fourier momentum, and s is the spin.

We now enumerate the degrees of freedom that the Fourier coefficients represent. The $2^{r/2-1}$ spins yield $2^{r/2-1}$ b's, which being complex due to $b^\dagger$'s give a total of $2^{r/2}$ real-valued degrees of freedom due to the b's. The d's similarly number $2^{r/2}$. Together there are

$$2^{r/2+1} \text{ real-valued degrees of freedom.} \tag{3.5}$$

The b's and d's have anticommutation relations that parallel the structure of wedge products described in section 2.8.[9]

Antisymmetry:

$$\begin{aligned} b(s)b(s') &= -\, b(s')b(s) \\ b^\dagger(s)b^\dagger(s') &= -\, b^\dagger(s')b^\dagger(s) \\ \\ d(s)d(s') &= -\, d(s')d(s) \\ d^\dagger(s)d^\dagger(s') &= -\, d^\dagger(s')d^\dagger(s) \\ \\ b(s)d(s') &= -\, d(s')b(s) \\ b^\dagger(s)d^\dagger(s') &= -\, d^\dagger(s')b^\dagger(s) \end{aligned} \tag{3.6}$$

Repeated factors are zero: any product of factors containing repeated factors are zero (Pauli) when applied to states

$$\begin{aligned} b(s)b(s) &\equiv 0 \\ d(s)d(s) &\equiv 0 \\ b^\dagger(s)b^\dagger(s) &\equiv 0 \\ d^\dagger(s)d^\dagger(s) &\equiv 0 \end{aligned} \tag{3.7}$$

Thus independent wedge products and *independent* creation/annihilation products have the same antisymmetric algebra.

We now turn to our PseudoQuantum fermions with two fields. We have pointed out their benefits in previous work.

3.3 PseudoQuantum Fermion Fields' Operators

If we consider the form of a fermion's PseudoQuantum fields[10] the number of b's and d's doubles

$$\begin{aligned} \psi_1(x) &= \Sigma_{\alpha,s}[b_1(\alpha, s)u_{\alpha s}f_\alpha(x) + d^\dagger_1(\alpha, s)v_{\alpha s}f_\alpha{}^*(x)] \\ \psi_2(x) &= \Sigma_{\alpha,s}[b_2(\alpha, s)u_{\alpha s}f_\alpha(x) + d^\dagger_2(\alpha, s)v_{\alpha s}f_\alpha{}^*(x)] \end{aligned} \tag{3.8}$$

---

9 For Fourier index α. It is understood that the creation and annihilation operators are explicitly or implicitly functions of momentum in this book.

10 In our PseudoQuantum formulation of fermions in the 1970's and recently we define two wave functions for each fermion for important reasoms presented in my earlier papers and books.

where α represents the Fourier momentum, then the total number of creation and annihilation (CAO) operators doubles to

$$d_{cr} = 2^{r/2 + 2} \tag{3.9}$$

which represent real-valued degrees of freedom.

The column length d_{cr} (used in Cosmos Theory) specifies degrees of freedom since the associated creation and annihilation operators generate independent states.

Note the set of PseudoQuantum type 1 operators b_1's, $b_1^\dagger$'s, d_1's, and $d_1^\dagger$'s all anticommute just as the independent asymmetric k-forms all anti-commute.

Antisymmetric products of creation and annihilation operators may be formed. Any product with repeated operators is zero when they are applied to the vacuum state. This limits the number of factors in a product to d_{cr} just as the k-forms are limited to have at most r factors in dimension r. Due to the algebraic properties[11] we find that there are

$$2^{R/2+1} = \text{total number of non-zero products of } b_1\text{'s, } b_1^\dagger\text{'s, } d_1\text{'s, and } d_1^\dagger\text{'s}$$
$$= 2^S \tag{3.10}$$

where $R = 2^{r/2+1}$ is the number of b_1's, $b_1^\dagger$'s, d_1's, and $d_1^\dagger$'s, and where $S = 2^{r/2} + 1$ using eq. 3.5.

The above comments also apply directly to the set of PseudoQuantum type 2 b_2's, $b_2^\dagger$'s, d_2's, and $d_2^\dagger$'s. The resulting total number of degrees of freedom are given by eq. 3.9.

The total number of non-zero products[12] of b_1's, $b_1^\dagger$'s, d_1's, $d_1^\dagger$'s b_2's, $b_2^\dagger$'s, d_2's, and $d_2^\dagger$'s – before reducing the set of products using the non-zero anti-commutators $\{b_1, d_2^\dagger\}$, $\{b_1^\dagger, d_2\}$, $\{b_2, d_1^\dagger\}$, $\{b_2^\dagger, d_{21}\}$ – is

$$2^{R'/2+1} = 2^{S'} \tag{3.11}$$

where $R' = 2^{r/2+2}$ is the number of all b's, $b^\dagger$'s, d_1's, and $d^\dagger$'s, and where

$$S' = 2^{r/2 + 1} + 1 \tag{3.12}$$

3.4 Dimension Arrays and Internal Symmetries

The column length d_{cr} is due to spin counting of a PseudoQuantum fermion. We now make the *ansatz* that there is an equal number of internal symmetry states represented by an index k. Then eq. 3.8 becomes

$$\psi_{1k}(x) = \Sigma_{\alpha,s}[b_{1k}(\alpha, s)u_{\alpha s}f_\alpha(x) + d^\dagger{}_{1k}(\alpha, s)v_{\alpha s}f_\alpha{}^*(x)] \tag{3.13}$$
$$\psi_{2k}(x) = \Sigma_{\alpha,s}[b_{2k}(\alpha, s)u_{\alpha s}f_\alpha(x) + d^\dagger{}_{2k}(\alpha, s)v_{\alpha s}f_\alpha{}^*(x)]$$

---

[11] The possible products may be enumerated by treating the set as a set of k-forms.

[12] These possible products may also be enumerated by treating the set as a set of k-forms.

The resulting total number of degrees of freedom (still respecting eqs. 3.6 and 3.7) is

$$d_{dr} = 2^{r/2+2}2^{r/2+2} = 2^{r+4} \quad (3.14)$$

3.4.1 Internal Symmetries Factor Justification

The introduction of the internal symmetries factor increases the number of degrees of freedom for a space. The number of creation and annihilation operators specifies degrees of freedom primarily associated with fermion spin. Similarly the internal symmetries factor specifies the number of degrees of freedom associated with the set of internal symmetry groups. One must chose the range of internal symmetries index in eq. 3.10. The only apparent *a priori* choice for the range of internal symmetry k index in eq. 3.10 is

$$2^{r/2+2} \quad (3.15)$$

This choice puts the degrees of freedom for the spin part of creation and annihilation operators on the same level as the degrees of freedom for internal symmetries. It gives an encouraging tightness of formulation. The choice is vindicated by the match between the r = 4 dimension array and the particle spectrums and set of internal symmetries of our Unified SuperStandard Theory.[13]

3.4.2 Dimension Array Features

The dimension r determines the number of components in a dimension array. *The dimensions within a dimension array are degrees of freedom.* Dimension arrays define the spaces of Cosmos Theory as shown in Fig. 3.1. As pointed out in previous books this number r plays two roles: 1) the overall dimension of the space and 2) within a dimension array it specifies the number of space-time dimensions of the space.

Having determined the dimension array for each even dimension r we may then proceed to allocate the fundamental fermion spectrum and the set of internal symmetries as well as other particle spectrums. In our four dimension universe our Unified SuperStandard Theory (UST) does this in detail. We have also specified the fermion spectrum and internal symmetries for higher dimension spaces as well.

The quadrupling pattern appearing in dimension arrays as the dimension r increases by two makes the determination of the form of higher dimension space-times easier. See chapter 4.

The form of dimension arrays, which is ultimately based on the number of independent antisymmetric tensors, is determined in the same manner as the form of γ-matrices.

3.5 The Nature of Dimension Arrays

The above derivation of γ-matrix form and dimension array form places them both as central to an understanding of elementary particles and space-times in our

---

[13] This match was the cause of our efforts in the study of hypercomplex spaces in the past five years beginning in Fall, 2019.

universe and other universes of higher dimension. It creates a basis for a truly unified theory of Nature, which our earlier books in the past four years demonstrate. It also provided a procedure in our recent books for the generation of the particles and symmetries of a space by General Relativistic-Internal Symmetry transformations from a single dimension in a Fundamental Reference Frame. Total Unification is achieved in 42 dimension and 88 dimension spaces.

COSMOS SPACES SPECTRUM

Blaha Space Number $N = o_s$	Cayley-Dickson Number n	Cayley Number d_c	Dimension Array column length d_{cn}	Dimension Array Size d_{dn}	Space-time-Dimension r
0	10	1024	2048	2048^2	18
1	9	512	1024	1024^2	16
2	8	256	512	512^2	14
3	7	128	256	256^2	12
4	6	64	128	128^2	10
5	5	32	64	64^2	8
6	4	16	32	32^2	6
7	**3**	**8**	**16**	$\mathbf{16^2}$	**4**
8	2	4	8	8^2	2
9	1	2	4	4^2	0
Limos :					
10	0	1	2	2^2	-2
11	-1	½	1	1	-4
12	-2	¼	½	$½^2$	-6
13	-3	1/8	¼	$¼^2$	-8
14	-4	1/16	1/8	$1/8^2$	-10

•
•
•

HYPERCOSMOS OF THE SECOND KIND SPACES SPECTRUM

Blaha Space Number $N = O_s$	Cayley-Dickson Number n	Cayley Number d_c	Dimension Array size d_{dN2}	Space-time-Dimension r	CASe Group $su(2^{r/2}, 2^{r/2})$ CASe
0	10	1024	1024×2048	18	su(512,512)
1	9	512	512×1024	16	su(256,256)
2	8	256	256×512	14	su(128,128)
3	7	128	128×256	12	su(64,64)
4	6	64	64×128	10	su(32,32)
5	5	32	32×64	8	su(16,16)
6	4	16	16×32	6	su(8,8)
7	**3**	**8**	$\mathbf{8 \times 16}$	**4**	**su(4,4)**
8	**2**	4	4×8	2	su(2,2)
9	**1**	2	2×4	0	su(1,1)
10	0	1	1×2	-2	
11	-2	½	½	-4	

Figure 3.1. The HyperCosmos, Limos, and the HyperCosmos of the Second Kind, space spectrums. Note r = 2n – 2.

4. Internal Quadrupling Structures

The n-forms, independent asymmetric tensors, γ-matrices, dimension arrays and Cosmos fractal curve all display a common internal quadrupling feature. We describe these features in this chapter. In general, when the dimension r, or the order n, increases by two for even r, or for order n, then the structure is quadrupled.

4.1 n-forms Quadrupling

When the even dimension r increases by two, the size 2^{n+2} of the resulting (n+2)-form is four times the size of the n-form 2^n. This is shown by example in section 2.8.1 with the detail described by eqs. 2.3 and 2.3a in particular.

$$T_n = T_{n-2} + 2\ T_{n-2}(1) + T_{n-2}(2) \tag{2.3}$$

4.2 Asymmetric Tensor Quadrupling

For independent asymmetric tensors there is also a factor of four increase when the even dimension r increases by two, the size 2^{r+2} of the resulting set of dimension r+2 tensors is four times the size of the set of dimension r tensors 2^r by eq. 2.6.

4.3 γ-matrices Quadrupling

The γ-matrices that appear in the form of $\gamma^\mu\partial_\mu$ in fermion Dirac equations also display a quadrupling for even dimensions r when the dimension is increased by two. The number of γ^μ is r in dimension r. The matrices have $2^{r/2}$ by $2^{r/2}$ rows and columns. Fig. 4.1 illustrates the quadrupling effect for r = 4 (our universe) and for r = 6 (the Megaverse).

Case r = 2

There are three Pauli matrices. These matrices appear in the r = 4 case.

Case r = 4

The matrices exhibit a nested structure quadrupling as the dimension increases by 2. The r = 4 γ-matrices contain r = 2 Pauli matrices. The three Pauli matrices in the upper right and lower left quadrants for γ^i where i = 1, 2, 3. The Dirac γ^0–matrix, where $\gamma^4 \equiv \gamma^0 = \text{diag}(1, 1, -1, -1)$, connects the upper and lower parts of these γ-matrices. It is the analogue of the Connection group interactions of Cosmos Theory.[14] Connection groups connect the parts of the Cosmos particle spectrums just as γ^0 connects the parts of Dirac matrices. The quadrupling effect is evident. The connection γ^0 matrix ties the parts of Dirac matrices together. The connection γ^0 is not displayed in Fig. 4.1.

---

[14] For Connection Groups see Appendix 9-B of Blaha (2023d). Connection Groups also appear in earlier books by the author.

Case r = 6

In fig. 4.1 the r = 6 diagram shows the form of four r = 6 Dirac matrices γ^μ where $\mu = 0, 1, 2, 3$. In the four r = 6 γ-matrices, there are r = 4 γ^k matrices where k = 0, 1, 2, 3. Each γ^k (where k = μ) appears twice in each γ^μ matrix in opposing quadrants. For example, the r = 6 matrix for γ^3 is

$$\gamma^3 = \begin{pmatrix} 0 & \text{DIRAC } \gamma^3 \\ \text{DIRAC } -\gamma^3 & 0 \end{pmatrix}$$

The number of these $2^{r/2}$ by $2^{r/2}$ γ-matrices in dimension r = 6 is r. We define two additional diagonal γ-matrices which we tentatively suggest may have a form analogous to the r = 4 case. Thus for r = 6 we take the additional two γ-matrices to be:[15]

$$\begin{aligned} \gamma^4 &= \text{diag}(1, 1, -1, -1, \ -1, -1, 1, 1) \\ \gamma^5 &= \text{diag}(1, 1, 1, 1, \ -1, -1, -1, -1) \end{aligned} \qquad (4.1)$$

They are not are not shown in Fig. 4.1. They correspond to the Connection groups of Cosmos Theory dimension arrays. They connect the γ-matrix parts in the four quadrants.

Case r > 6

For even dimensions r > 6 we define $\gamma_r{}^k$ as the kth γ-matrix in dimension r:

$$\gamma_{r+2}{}^k = \begin{pmatrix} 0 & \gamma_r{}^k \\ -\gamma_r{}^k & 0 \end{pmatrix}$$

for k = 1, 2, …, r – 1.

---

[15] Representations of these γ-matrices are equivalent up to a unitary transformation by Good's Theorem generalized to higher dimensions. See R. H. Good Jr., Rev. of Mod. Phys., **27**, 187 (1955).

In r + 2 dimensions, there are r γ-matrices $\gamma_{r+2}{}^{\mu}$ where μ = 0, 1, 2, … , r – 1. Within these γ-matrices, there are $\gamma_r{}^k$ matrices where k = 0, 1, 2, … , r – 1. Each such $\gamma_r{}^k$ appears twice in each $\gamma_{r+2}{}^k$ matrix in opposing quadrants. These matrices total in number to r.

We define two additional diagonal γ-matrices which we tentatively suggest may have a form similar to the r = 4 case:

$$\gamma_{r+2}{}^{r} = \text{diag}(\gamma_r{}^{r}, -\gamma_r{}^{r}) \qquad (4.2)$$
$$\gamma_{r+2}{}^{r+1} = \text{diag}(1, 1, 1, 1, \ldots, -1, -1, -1, -1, \ldots)$$

where the number of 1's and -1's each have $2^{r/2}$ entries in $\gamma_{r+2}{}^{r+1}$. The number of entries total to $2^{(r+2)/2}$. The full set consists of (r + 2) $\gamma_{r+2}{}^{\mu}$ matrices where μ = 0, 1, 2, … , r + 1.

Thus there is γ-matrix quadrupling with a pair of additional diagonal connection matrices added each time as r increases by 2.

4.4 Dimension Array Quadrupling

Dimension arrays quadruple in size as the dimension r increases by 2 due to eq. 3.11. Figs. 4.2 and 4.3 show the quadrupling effect, which nicely corresponds to the structure of the fermion spectrum.

The nesting quadrupling property carries through for the fermion spectrum and group symmetry arrays. Figs. 5.3 and 5.4 show the nested fermion structure.

The nesting of the group symmetries is modified after a quadrupling to connect various quarters together. Some of the symmetry groups are defined to connect quarters together using Connection Groups that we have described in earlier books such as Blaha (2023d). Connection groups parallel the connection γ-matrices described above that appear as the dimension r increases by two.

If one considers the known Standard Model part within these diagrams one sees it fits in the nested dimension arrays like the pieces in a puzzle. Thus there is experimental support for Cosmos Theory structuring in The Standard Model.

See Blaha (2023d) *Newton's Apple is Now the Fermion* and (2023e) *Cosmos Theory: The Sub-Particle Gambol Model* as well as our earlier books for additional detail.

4.5 Hilbert and Cosmos Fractal Curves Quadrupling

The quadrupling seen in γ-matrices and dimension arrays is paralleled in the form of Hilbert fractal curve and Fractal Cosmos[16] curve. These curves are very similar. The primary difference is Cosmos Dimension Array Column Length is twice the Hilbert Curve Line Length order by order of curve construction where the Dimension Array Cayley-Dickson number $n = n_H - 1$ with n_H being the Hilbert curve construction order.

[16] We use the name Cosmos Curve in preference to Cosmic Curve.

4.5.1 The Fractal Cosmos Curve

We now use the fractal relations first found in Blaha (2024b). The fractal construction of the Cosmos Curve is based on the association of dimension array column lengths[17] and the orders of piecewise linear line segment lengths that are used to form a fractal curve. Eq. 4.3 relates the *Cosmos Curve* construction to the Hilbert curve construction.[18]

Cosmos Cayley-Dickson number $n = n_H - 1$ (4.3)
Hilbert Curve Line Length $= 2^{n_H}$
Hilbert Number of "boxes" $= 2^{2n_H}$
Cosmos Dimension Array Column Length $= 2^{n+1} = 2^{n_H}$
Cosmos Dimension Array Size (number of elements in array) $= 2^{2n+2} = 2^{2n_H}$

where n is the Cosmos Cayley-Dickson number and n_H is the order in the Hilbert curve construction. A correspondence may be set up between Cosmos Dimension array diagrams and Hilbert curve line segment diagrams. Figs. 4.4 and 4.5 show the map between Cosmos Curve diagrams and equivalent Hilbert curve diagrams.

Non-negative n (or r) dimension array column lengths combine to generate a two dimension filled square fractal grid from a one dimension line segment.

The negative n (r) dimension array column lengths for n = -2 through n = $-\infty$ combine to form a one dimension line element of length 1 equal to the array column length 1 for n = -1 from a zero dimension point due to the identity

$$\sum_{n=1}^{\infty} 2^{-n} = 1 \qquad (4.4)$$

If we adjoin the dimension array column lengths for $n = -\infty$ through $n = \infty$ then we have the *Cosmos Curve from a zero dimension point to a two dimension filled square grid.*

The fractal dimension generation is paralleled by the generation of dimension arrays for spaces at each step in its construction.

4.5.2 The Two Dimension Square Grid Cosmos Curve

The Cosmos Curve generates a two dimension filled square grid. This grid might be viewed as the beginning of a new Cosmos spaces spectrum. It embodies the 4 dimensions of the 2×2 n = 0 dimension array d_{d0}, which is the precursor of the ten HyperCosmos spaces. It may be viewed as a continuous grid corresponding to d_{d0}.

The four corners of the Cosmos Curve square grid are mapped to the four dimensions of an n = 0 dimension array. We can associate that dimension array with the $n_H = 1$ Hilbert order diagram and proceed to construct a new Hilbert-like curve and new associated sets of Cosmos dimension arrays from it. At each order of construction we

---

[17] Order by order (dimension by dimension) the dimension array column lengths map to the length of the corresponding Hilbert curve piecewise line lengths. Blaha (2024b).

[18] The construction of the fractal curve corresponding to Cosmos Theory spaces was shown to be similar to the construction of the Hilbert feactal curve in Blaha (2024b).

take the boxes of type C in Fig. 4.4 and transform them to boxes of type D filling each with a "color pattern." As the order n_H increases, an irregularly shaped figure is generated that becomes a set of interconnected islands of "Cosmos" color.

This new construction of interconnected islands might be called a *SuperCosmos* archipelago. It corresponds order by order to supersets of Cosmos spaces with filled grids used to introduce continuity between dimensions in each 4 by 4 block. Thus a continuum is established between the elements of each diagram. The elements are, of course, Cosmos Theory sets of spaces.

4.5.3 Comparison between Fractal Formulation and Creation/Annihilation Operator Formulation

The fractal formulation, and the creation/annihilation operator based formulation, offer somewhat different approaches to Cosmos Theory. The fractal formulation, which has a Hilbert-like curve view of the Cosmos Curve, offers perhaps a simpler, more elegant formulation. The fractal formulation has the same nesting as the Cosmos Theory dimension arrays.

The fractal formulation does not have some significant features of the Creation And Annihilation Operator (CAA) formulation :

1. ***The basis of the CAA formulation in the number of independent tensors singles out the CAA formulation as fundamental.***

2. The choice of even number space-time dimensions only is natural in the CAA formulation since the total number of creation and annihilation operators in an odd dimension space equals the number in the next lower even dimension – thus giving a potential redundancy that is removed by choosing even dimensions only.

3. The CAA formulation creates a dimension array that has extended General Relativistic – Internal Symmetry transformations that parallel the corresponding transformations of the creation/annihilation operators of fermion fields.in General Relativity.

4. The CAA formulation allows one to define a Fundamental Reference Frame (FRF) that can map a single primordial dimension to complete dimension arrays, which introduces a fundamental simplicity into the Cosmos.

5. Space-time dimensions may be extracted from a dimension array.

6. Manipulation of the quantum fields based on the CAA formalism enables the creation of totally unified quantum field theories through the generation of 42 and 88 space-time dimension spaces. See Figs. 3.6 and 4.7.

We conclude the CAA tensor-based formalism is another basis for Cosmos Theory and the UST for our universe.

4.5.4 The Cosmos Curve and Unification and Fundamental Reference Frame (FRF)

In Blaha (2023a) we developed a unification of the ten HyperCosmos spaces and the 10 Second Kind HyperCosmos in a 42 space-time dimension Full HyperUnification space. See Fig. 3.1. The purpose was to create a space from these 20 spaces that supported a combined General Relativistic and symmetry group transformation that would generate the dimension arrays of all 20 spaces from one dimension located in a Fundamental Reference Frame (FRF).

In section 3.3.1 of Blaha (2023a) we described the nature of an FRF. A dimension array Reference Frame treats a dimension array with d_{dr} components as a vector. The vector is subject to a combined General Relativistic – Internal Symmetry transformation to produce a new dimension array in vector form. One may chose the Reference Frame to be an FRF where the vector has only one non-zero dimension. This FRF is analogous to the rest frame of a particle with only one non-zero momentum component: its energy equals its mass. The analogy to Special Relativistic transformations from a rest frame to a moving frame is clear.

Using FRF's allows the dimension array of any space to be reduced to one non-zero component.

An examination of Fig. 4.6 shows that the 10 dimension arrays' column lengths and 10 Second Kind HyperCosmos dimension arrays' column lengths combine to form the 42 space-time dimension Full HyperUnification space shown in Fig. 4.7. The column lengths of the HyperCosmos part are a subset of the column lengths in the Cosmos Curve. See Blaha (2023a) for details.

The dimension array of the 42 space-time dimension Full HyperUnification space can be treated as a vector in an 88 space-time dimension UltraUnification (UU) space. Then an FRF may be defined for it with only one non-zero dimension that enables the UU dimension array to be generated from the FRF with only one non-zero dimension.

Thus the UU enables a reduction of all Cosmos Theory dimension arrays to one primordial non-zero dimension – an ultimate form of unification of all Cosmos Theory spaces. See Fig. 4.7 for the structure of the complete set of Cosmos spaces. There are four levels in the Cosmos, which is the most economical approach to generating the Cosmos. The one primordial dimension in UU can be fractionated into an infinite dust of fractionated dimensions using Limos.

The spaces of Cosmos Theory (Fig. 4.7) *are*

!0 HyperCosmos spaces + their 10 HyperUnification spaces + (4.5)
+ 10 Second Kind HyperCosmos spaces +
+ the 10 Second Kind HyperUnification spaces +
+ the 42 Dimension Full HyperUnification Space +
+ the 88 Dimension UU space = 42 spaces

+ Limos[19] (not shown in Fig. 4.7)

4.5.5 The Language of the Cosmos Curve

The Hilbert curve and thus the Cosmos Curve may be viewed as generated according to a language ((L System). The language specification is

Alphabet: α, β
Constants: φ + –
Axiom: α

Production Rules:
α → +βφ–αφα – φβ+
β→ –αφ+βφβ+φα-

The language interpretation of the Hilbert and Cosmos Curves introduces an interesting new view of Cosmos Theory raising the issue of a Cosmos Language. Languages have been associated with Physics previously. See Blaha (1998) and (2005c) for detailed examples.

4.6 Analogous γ-Matrix – Dimension Array Features

The even space-time dimension γ-matrices exhibit properties similar to the Cosmos Theory dimension arrays:

1. Dimension arrays have a size that is 16 times the size of the corresponding γ-matrices.

2. The γ-matrices are square. They are r matrices of size $2^{r/2}$ by $2^{r/2}$. As the even space-time dimension r increases by 2 the γ-matrices row and column sizes double. The γ-matrices quadruple as a result. For example the r = 6 γ-matrices are 8 by 8 matrices containing a quadruple of r = 4 γ-matrix parts.

3. The γ-matrices exhibit a form of nesting in quadruples. Fig. 4.2 displays nesting from r =2 (Pauli matrices) to r = 4 and then to r = 6 quadrupling at each dimension increase by 2.

4. Fractional γ-matrices of *negative* dimension n may be defined by quartering the γ-matrices of the next higher space-time dimension. These fractional matrices may be of interest for gambol quantum field theory. This topic has been considered in our previous books.

[1919] Limos spaces with their negative space-time dimensions are not part of the Full HyperUnification space or the UltraUnification space because they do not participate in the unification of the twenty HyperCosmos spaces.

The Hilbert curve features, Cosmos dimension array features, and γ-matrix structural features are the same.

4.7 Future Directions

The fundamental fact of Cosmos Theory, the number of independent tensors in a space, and the assumptions that then lead directly to the Cosmos spaces spectrum, make Cosmos Theory the most likely theory of the Cosmos.

This conclusion motivates the following program for Elementary Particles:

1. Study other additional features of Cosmos Theory.

2. Perform experiments to expand the fermion spectrum to a fourth generation and to three additional layers. Find the Dark Matter part.

3. Explore the implications of the Generation and Layer groups – particularly for the expansion of the Cabibbo-Kobayashi-Maskawa (CKM) matrix.

4. Look for Astrophysical support for Cosmos Theory – particularly for the Megaverse and higher dimension universes.

The spectrum of particles and symmetries needs to be expanded from The Standard Model to the predictions of Cosmos Theory and the Unified SuperStandard Theory (UST).

Success in these endeavors will lead to a new Physics. The possibility of several new accelerators operating at 3 to 8 times current maximum energies is encouraging.

This author *solely* developed Cosmos Theory over the past five years beginning with The Unified SuperStandard Theory (UST) found in his previous work.

4.8 The Path of Investigation

The Standard Model →

Unified SuperStandard Theory →

Hypercomplex number based extensions →

Cosmos Theory →

Fractal Cosmos Curve Theory →

Tensor based Fractal Cosmos Theory

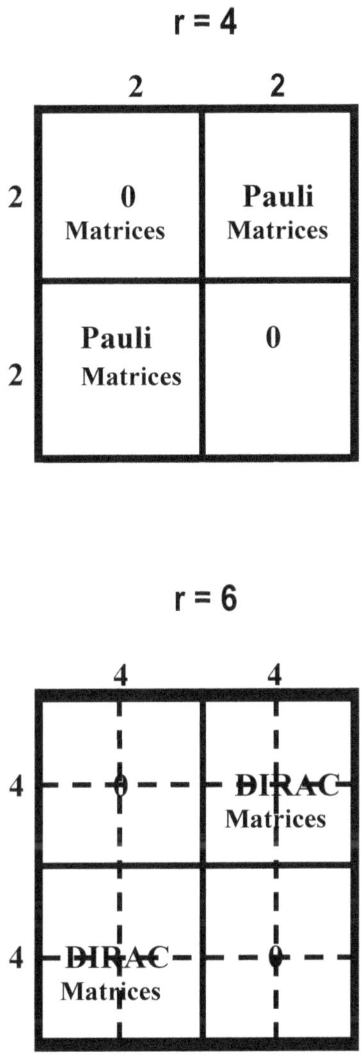

Figure 4.1. The γ-matrices for r = 4 and r = 6 space-time dimensions. The matrices exhibit a nested structure quadrupling as the dimension increases by 2. For example, the r = 4 γ-matrices contain the r = 2 Pauli matrices. The r = 4 diagram shows the three Pauli matrices in the upper right and lower left quadrants. The $\gamma^0 = \gamma^4 = \text{diag}(1, 1, -1, -1)$ matrix is not shown. The r = 6 diagram shows four Dirac matrices with each in a quadrant. There are two additional γ-matrices. See sections 4.3. and 4.4. The additional γ-matrices that are not shown correspond to the Connection groups of Cosmos Theory dimension arrays. They connect the parts in γ-matrix quadrants.

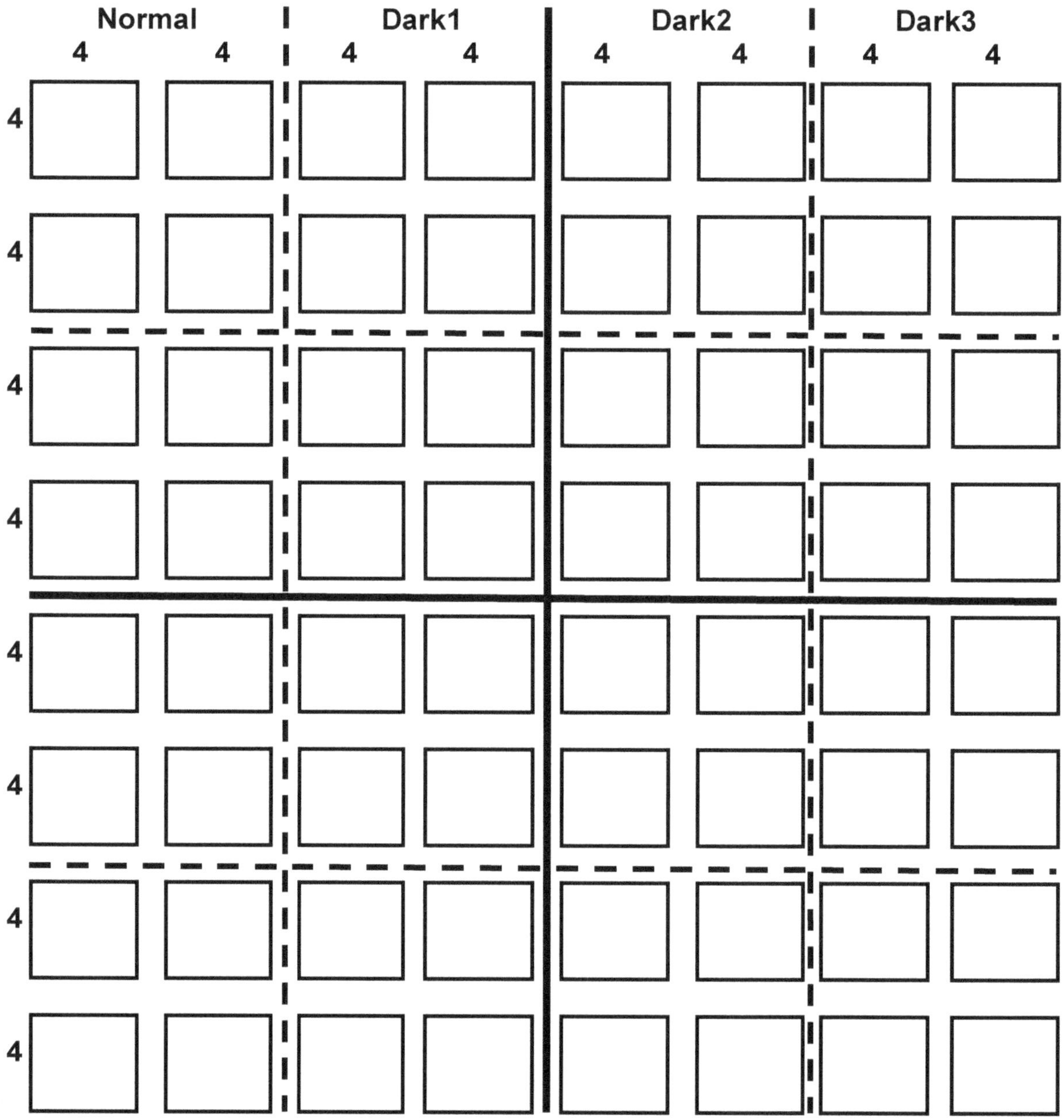

Figure 4.2. The n = 4 (where r = 6) Megaverse dimension array. Four quadruplings are visible. Each quadrant is an n = 3 (r = 4) UST dimension array. There is a visible substructure. There are sixty-four 4 × 4 n = 1 blocks. Each 8 × 8 quadrant is an n = 2 dimension array. Each 16 × 16 quadrant is an n = 3 dimension array. An exact match with Hilbert orders nesting is evident. This figure is from Blaha (2020d) for the Megaverse. Note r = 2n – 2. See Fig. 3.1.

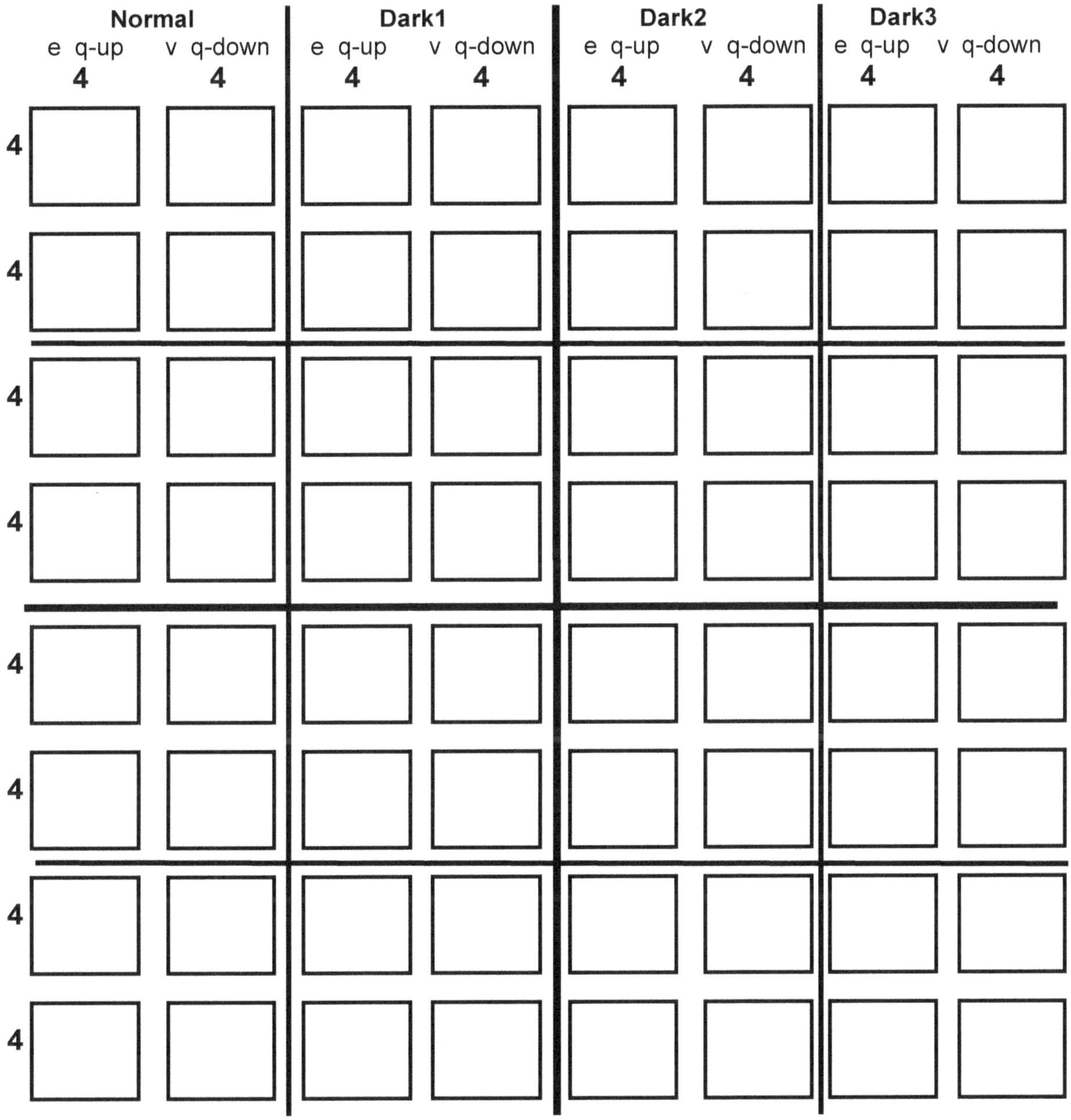

Figure 4.3. Partition of the fermion 32 × 32 Cayley-Dickson n = 4 (Megaverse) dimension array. Four quadruples are visible. There are sixty-four 4 × 4 n = 1 blocks. Each 8 × 8 quadrant is an n = 2 dimension array. Each 16 × 16 quadrant is an n = 3 dimension array. An exact match with Hilbert orders is evident. The label e q-up indicates a charged lepton – up-type quark pair. The label v q-down indicates a neutral lepton – down-type quark pair, and so on. This figure is from Blaha (2020i) for the UST and Megaverse. Note r = 2n – 2. See Fig. 3.1.

n_H = 1 Hilbert Line

A

Extend to make a Cosmos Cayley-Dickson n = 0 Dimension Array Box (of 4 dimensions)

B

Hilbert Order n_H = 2 Fractal View, a Cayley-Dickson n = 1 Dimension Array Box (of 16 dimensions)

C

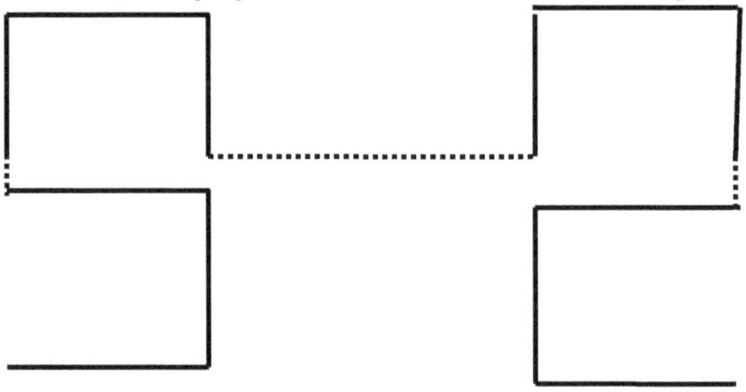

Hilbert Order n_H = 2 and n = 1 Cosmos View

D

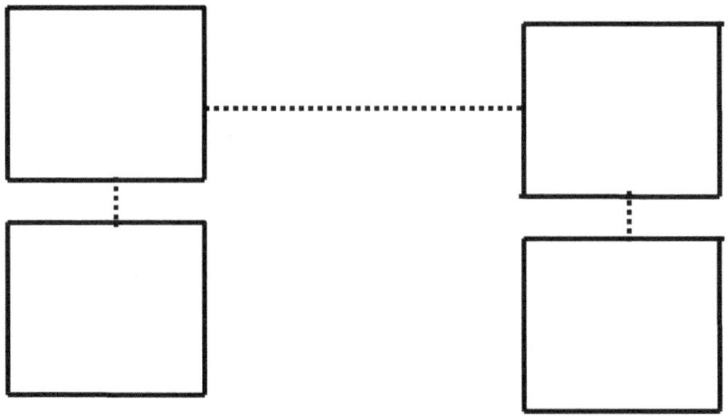

Cosmos Cayley-Dickson n = 1 Dimension Array Form (of 16 dimensions)

E

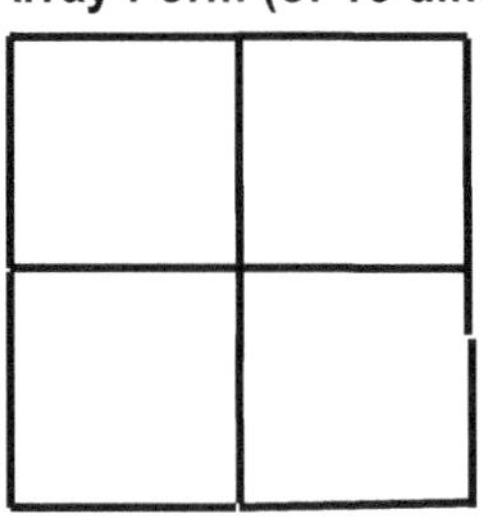

Figure 4.4. The transition from the Hilbert curve Fractal diagrams to the Cosmos Theory dimension array diagrams. Every Hilbert fractal diagram has an equivalent Cosmos Theory dimension array.

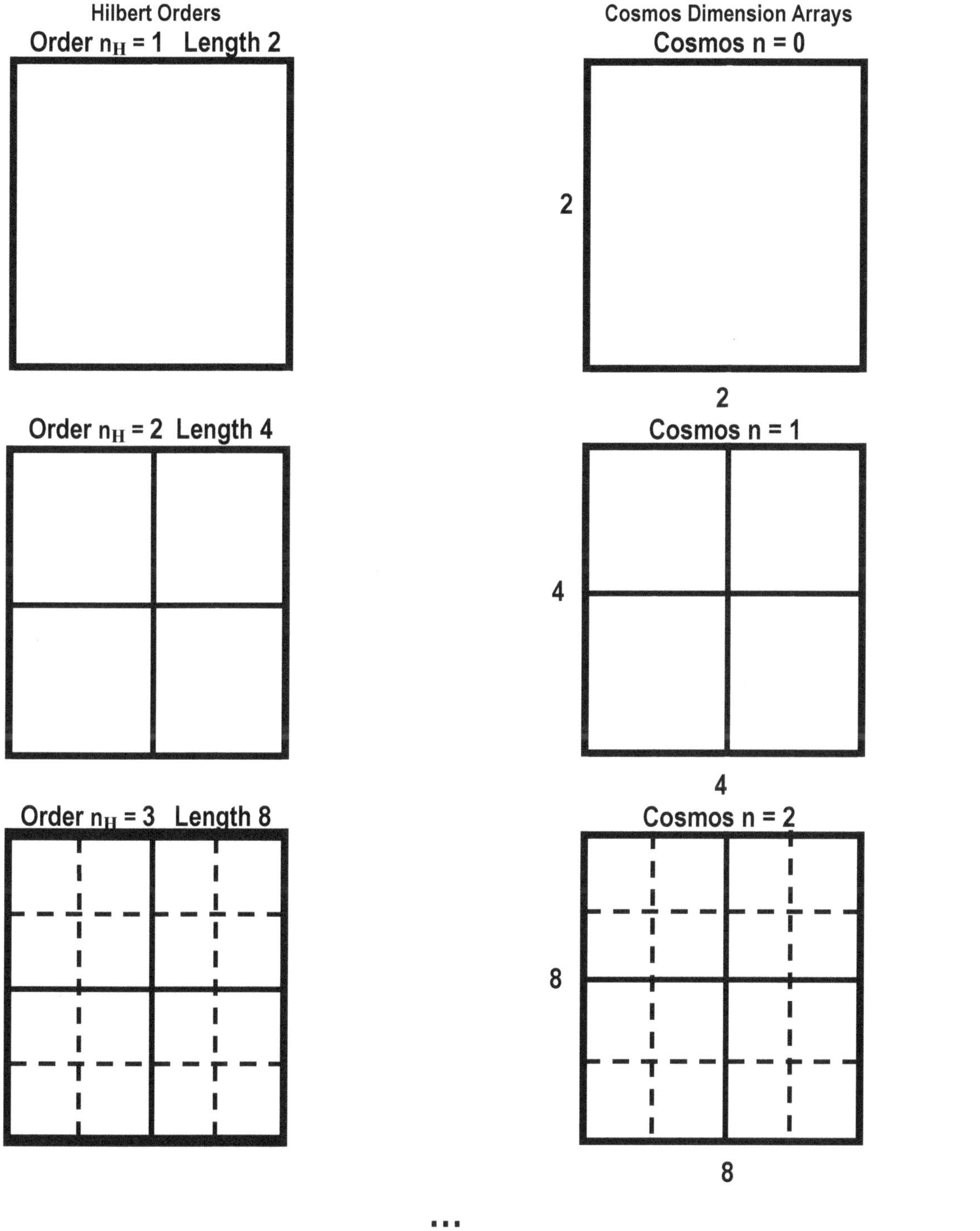

Figure 4.5. Low order Hilbert line segments depicted as nested sets of boxes as in Fig. 4.4 compared to corresponding Cosmos Theory dimension arrays ordered by Cayley-Dickson numbers. Dimension array sizes are displayed.

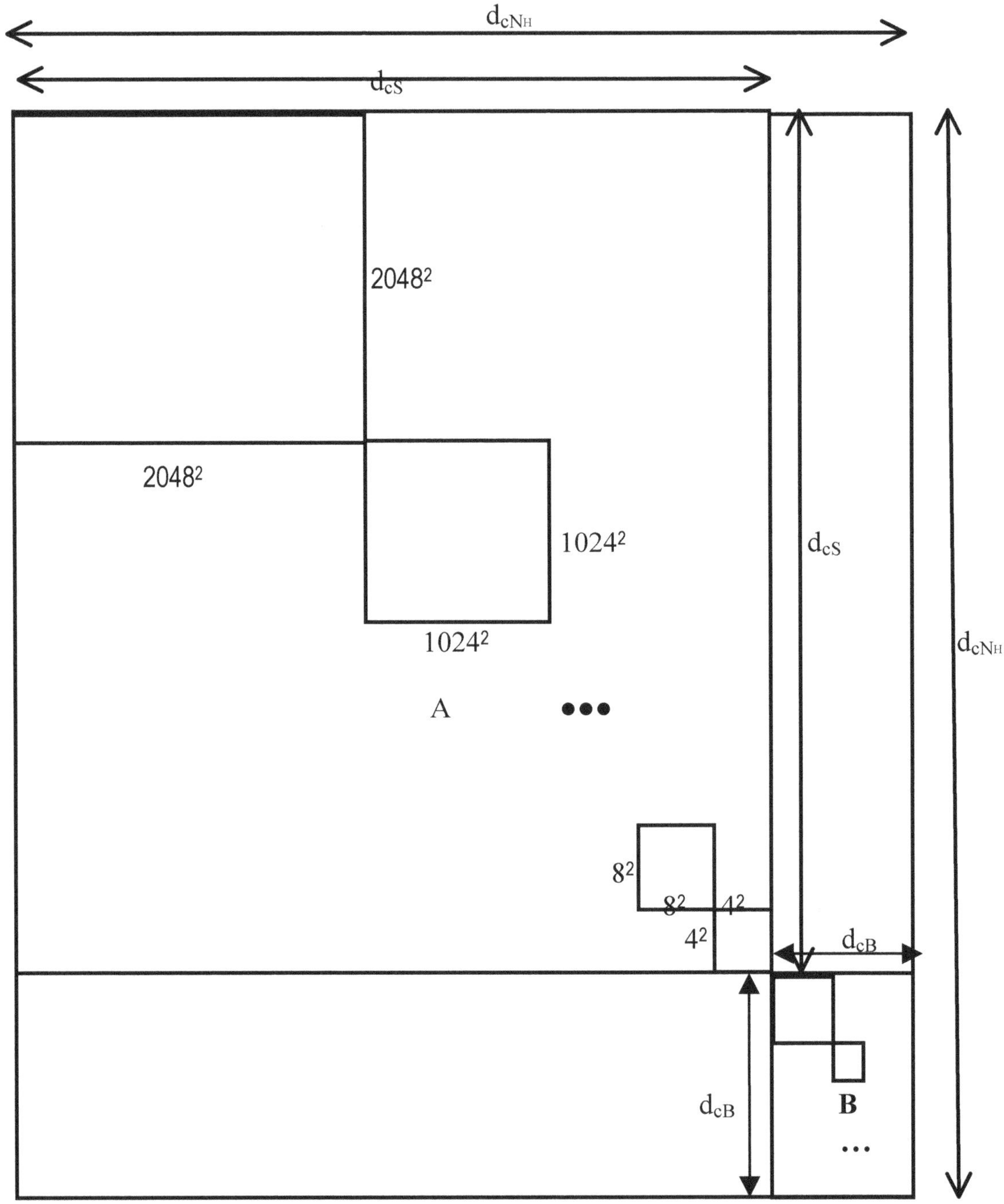

Figure 4.6. Form of a HyperUnification transformation. It is also the form of the dimension array of 42 dimension space-time. Ten HyperCosmos dimension array's column lengths are shown. Ten Second Kind HyperCosmos dimension array's column lengths are contained in the d_{cB} by d_{cB} block. The figure is not drawn to scale.See Chapter 6 of Blaha (2023a) for details.

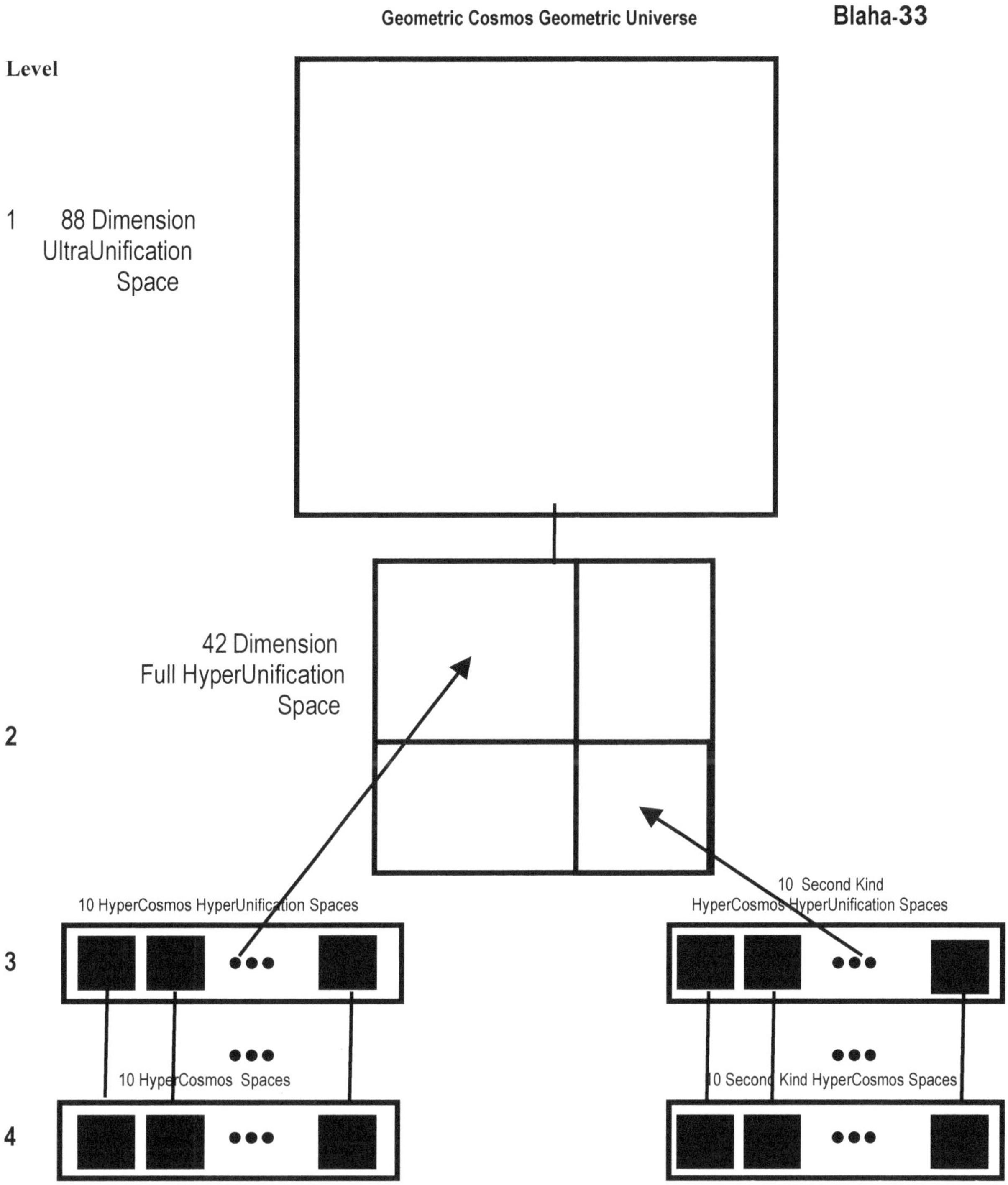

Figure 4.7. Diagram of the four levels of the Cosmos Spaces. They contain 42 spaces with no contributions from Limos spaces which are treated separately.

5. Dimension Array Decomposition

The dimension array for dimension r has

$$2^{r+4} \tag{5.1}$$

components. Within it are

$$2^{r+1} \tag{5.2}$$

parts consisting of 8 real-valued dimensions for one of the direct products

$$\begin{array}{l} SU(4) \\ SU(2)\otimes U(1)\otimes SL(2, \mathbf{C}) \\ U(4) \\ SU(2)\otimes U(1)\otimes SU(2)\otimes U(1) \end{array} \tag{5.3}$$

according to Figs. 5.1 and 5.2 below.

We now decompose dimension arrays into an array totaling 2^{r+1} elements. Each element consists of 8 real-valued dimensions.

$$d_{nr} = 8 \sum_{i=1}^{2^{r+1}} d_i = 2^{r+4} \tag{5.4}$$

where

$$d_i = \text{the } i^{th} \text{ element in the dimension array} \tag{5.5}$$

The γ-matrices appear in Dirac equations in the form $\gamma\bullet\partial$. Similarly we create an ordered set of internal symmetry irreducible representation matrices, T_{imj} ordered in the same way as the elements in d (for matrix indices m and j).

The d_i are sets of 8 real-valued dimensions for irreducible representations of internal symmetries. We may identify the d_i as representing parts of fermion wave functions (See section 5.1 for more details.) with internal symmetry indices k appearing in CAA operators in eq. 3.13 such as $b_{1k}(\alpha, s)$. Each d_i represents a set of values of k.

We can form internal symmetry representations.[20] The form of this combination of dimension array parts and internal symmetry matrices T_i may be represented by

$$T_{imj}d_i \tag{5.6}$$

---

[20] Note that the parts consist of direct products. Consequently the corresponding matrices must be treated similarly.

where the T_{imj} and d_i are formed from the respective representation internal symmetry matrices T_i irreducible representation, and the corresponding dimensions in d_i with i labeling the internal symmetry. We then form the products with vector boson wave functions $A_m{}^\mu$

$$A_i{}^\mu T_{imj} \tag{5.7}$$

to represent terms in corresponding fermion Dirac equations with d_i representing coordinates within fermion wave functions.[21]

Thus each set of dimension array parts, d_i, maps to the wave function indices of internal symmetries. A dimension array represents a set of internal symmetry dimensions that maps directly to internal symmetry matrices.

Similarly, the γ-matrices of a space separate into parts such as the individual γ^μ matrices. The similarity of γ-matrices, and dimension arrays combined with internal symmetry irreducible representation matrices, is evident.

5.1 Details of the Transition from the d_i to Interaction Terms

The definition of internal symmetry interaction terms based on the d_i parts of a dimension array is based on the general form of fermion wave functions (eq. 3.13):

$$\begin{array}{llcl}
k = 1 & b_{p1}(\alpha, s) & & d^{\dagger}{}_{p1}(\alpha, s) \\
\quad 2 & b_{p2}(\alpha, s) & & d^{\dagger}{}_{p2}(\alpha, s) \\
\ldots & & \ldots & \\
 & \multicolumn{3}{l}{\text{[a set of } d_{ij} \text{ dimension wave function terms for internal symmetry i]}} \\
\ldots & & \ldots & \\
\psi_{pk}(x) = & \ldots\, b_{pk}(\alpha, s) \ldots & + & d^{\dagger}{}_{pk}(\alpha, s) \;\ldots \\
\ldots & & \ldots & \\
2^{r+4} & b_{p2^{r+4}}(\alpha, s) & & d^{\dagger}{}_{p2^{r+4}}(\alpha, s)
\end{array} \tag{5.8}$$

where the internal symmetry index is k, where the last wave function above has internal symmetry index 2^{r+4}, and where p = 1, 2 is the PseudoQuantum index.

When an internal symmetry matrix T_i is combined with a wave function, then the internal symmetry indices of the matrix T_{imj} combine with the corresponding internal symmetry indices of the wave function:

$$\Sigma_j T_{imj}\psi_{2j} = \sum_{\substack{j \text{ contained} \\ \text{in } d_i}} T_{imj}\psi_{2j} \tag{5.9}$$

---

[21] We note again that the d_i represent subsets of a fermion's wave function's internal symmetry indices (eq. 3.13).

where 2 is the PseudoQuantum index (chosen arbitrarily) and i labels the internal symmetry with matrix indices m and j. Then the interaction term in field equations is

$$A_i^{\mu} T_{imj} \psi_{2j} \equiv A_i^{\mu} T_i \psi_2$$

5.2 Tentative Experimental Confirmation

An examination of Fig. 5.3 shows that the d_1 part of the dimension array is exactly The Standard Model. One expects the next part that will be found is d_2. They would appear in the analysis of the four generation version of the Cabibbo-Kobayashi-Maskawa (CKM) matrix.

Significant confirmation would involve discovery of fourth generation fermions, three additional layers of fermions, and the discovery of Dark matter fermions.

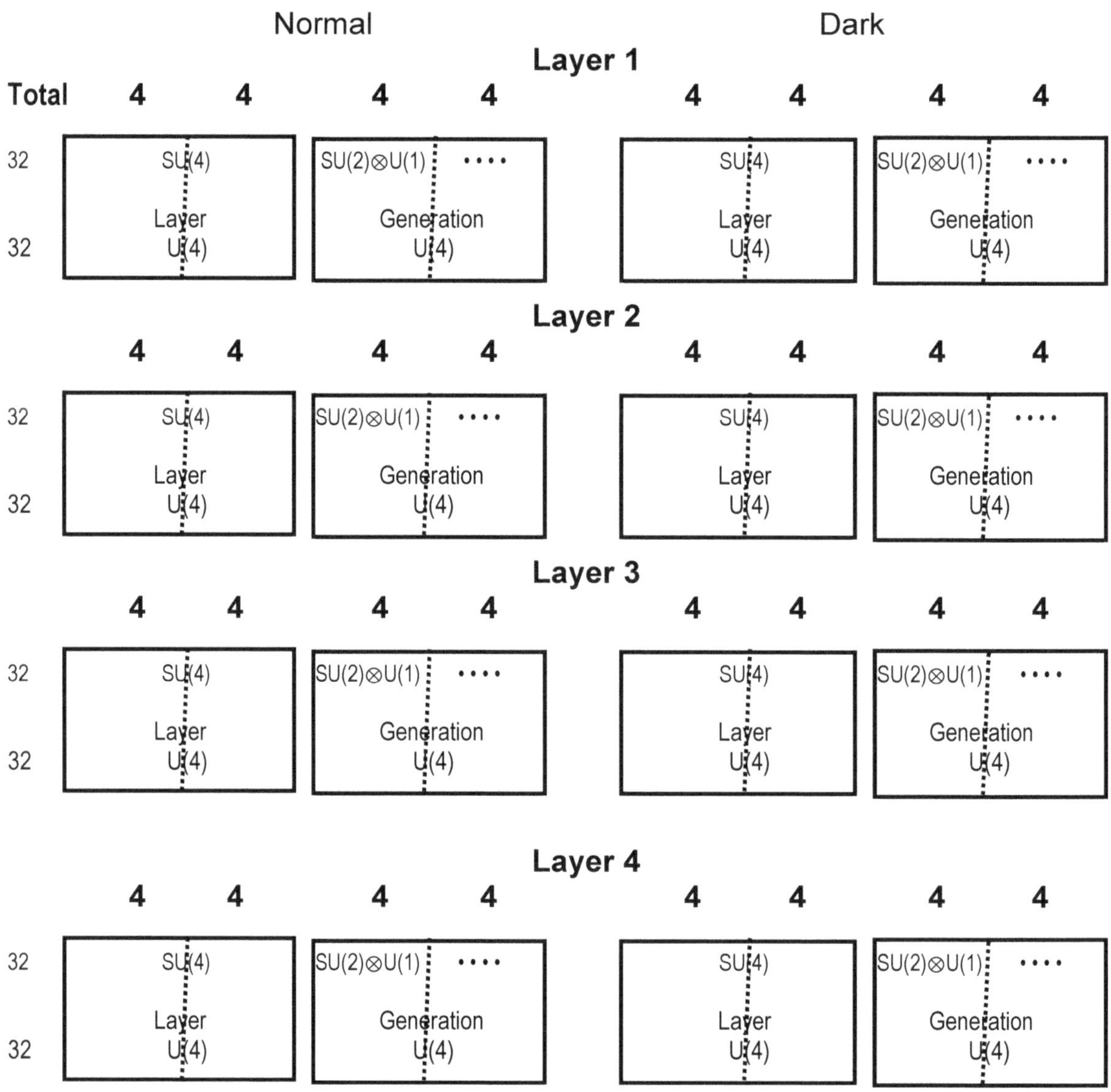

Figure 5.1. Schematic for the Normal and Dark symmetries of UST. The real and imaginary dimensions are counted separately. For example SU(4) has four real dimensions and four imaginary dimensions. Each layer has 32 + 32 = 64 dimensions. There are 256 dimensions corresponding to the 256 entries in the dimension array for UST where n = 3. The 32 • 's form sets of 4 dimensions were initially allocated to 4D space-times. These dots are mapped to SL(2, **C**) and 7 SU(2)⊗U(1) Connection groups. Three groups unite layers, and 4 groups unite corresponding Normal and Dark layers This figure is from Blaha (2020d) for the UST.

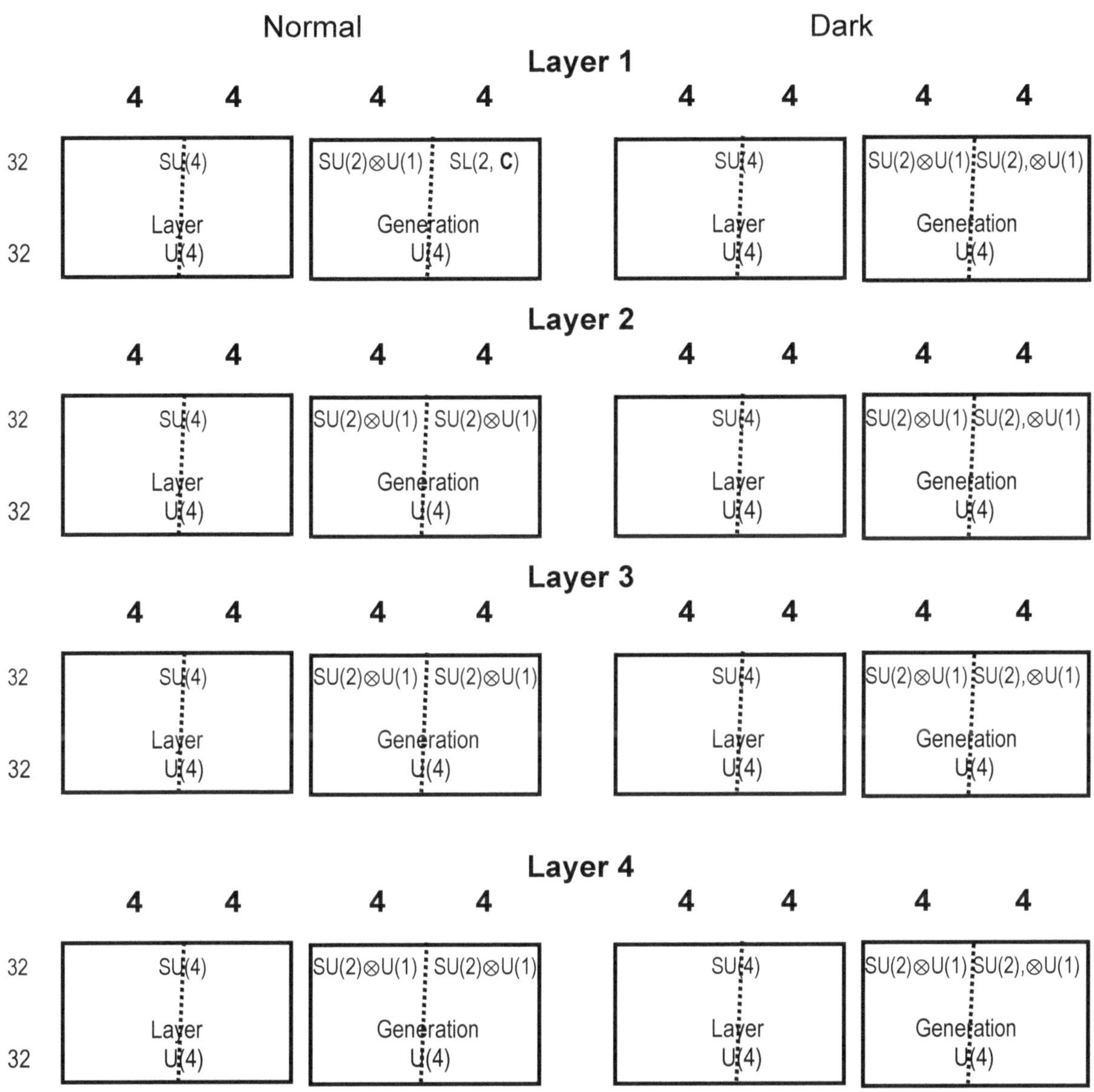

Figure 5.2. Schematic for the Normal and Dark symmetries of UST. Seven sets of dots in Fig. 5.1 are replaced by SU(2)⊗U(1) Connection groups. SL(2, C) represents the Lorentz group $SO^{+}(1,3)$. This figure is from Blaha (2020d) for the UST.

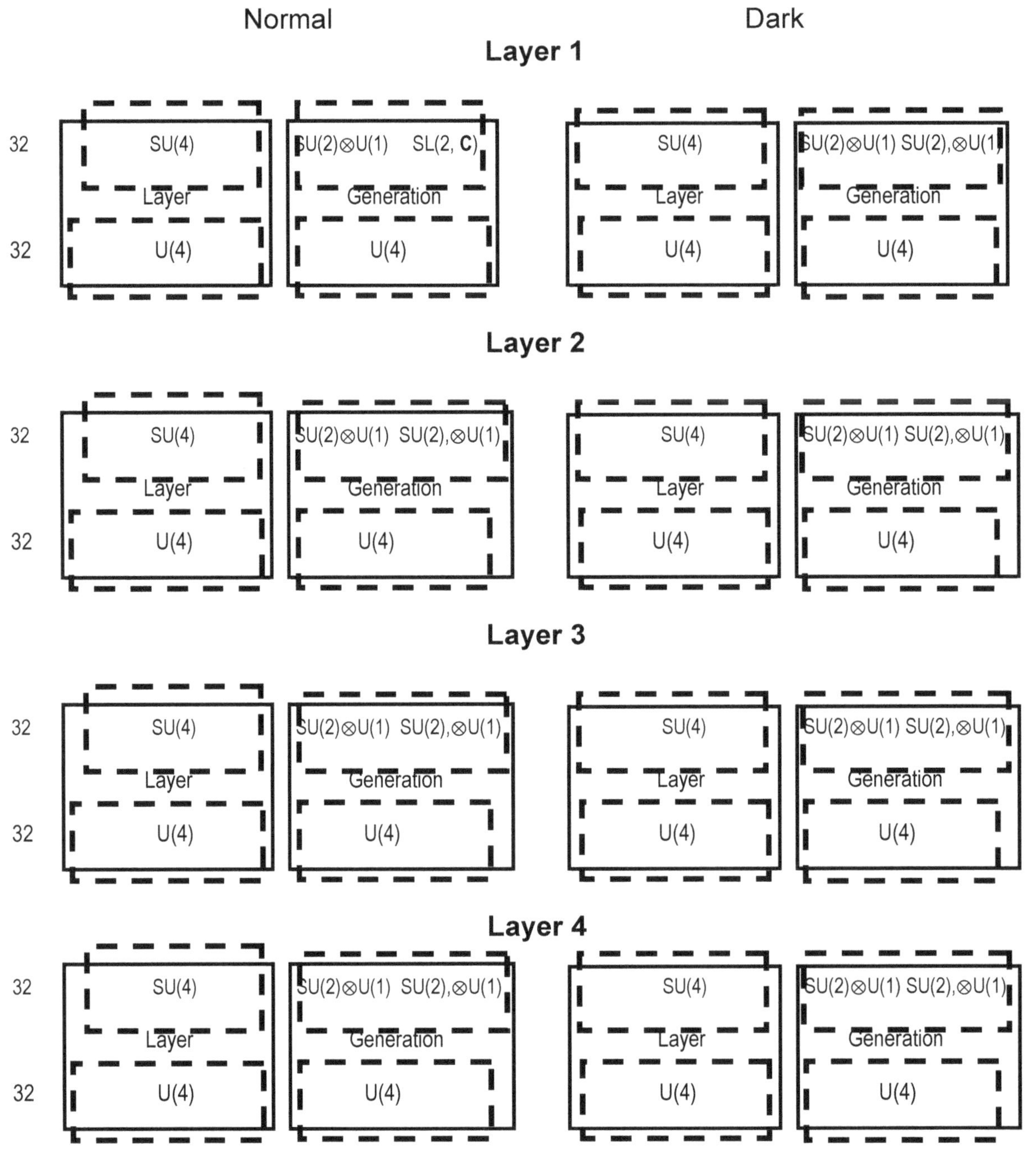

Figure 5.3. The thirty-two 8 dimension d_i parts of the r = 4 UST dimension array. The r = 6 dimension array (not shown) has 128 8 dimension d_i parts.

6. Cosmos Spaces and UST

Cosmos Theory has an r = 4 dimension space as the space of the UST. The UST has a subset corresponding to The Standard Model. Some points of difference are: UST has four layers, which each consist of four generations of eight fundamental fermions (leptons and quarks); UST has a matching Dark sector (although this may be eliminated easily); UST has multiple ElectroWeak and Strong interactions that differ in differing layers. These points are covered in more detail later and also in earlier books by the author.

The Cosmos Theory spectrum of spaces (Fig. 3.1) does not create universes. The spectrum is a template for the structure of universes. A space is a structure without substance. Universes must be defined/created separately. The creation of a universe requires an initial mass/energy. Universes are composed of substances. They begin as a state (A Big Bang state it seems) that unfolds according to a structure specified by the dimension array for its space-time dimension. A dimension specifies the universe's structure. Within the dimension array for a universe there are dimensions for internal symmetry groups and space-time dimensions. The space-time dimensions within a dimension array total to the dimension of the space.[22]

The sequence of events leading eventually to universes is

Define a sequence of integers n

For each n, one can define a set of independent asymmetric tensors

Generate a spectrum of spaces

Then derive the size of the dimension arrays for each space.

Use the dimension array to define internal symmetries' fundamental representations dimensions and space-time dimensions

Use the quadrupling feature to extrapolate from the internal symmetries to the structure of higher dimension spaces

Generate one or more universes

---

[22] Thus the 10 dimensions used to define the set of 10 Cosmos spaces also appear within the dimension arrays.

6.1 Generation of the Cosmos Universes[23]

The definition of the Cosmos spaces does not, in itself, specify any universes. A universe is a structure containing mass-energy. It must be generated. It uses a Cosmos space structure as a template. We have developed quantum field theories of universes with interactions in previous books. These theories are renormalizable in any dimension to any order in perturbation theory using the author's Two Tier formalism of Blaha (2002).

There are a number of ways to generate universes:

1. Generate all universes *ab initio.*

2. Generate the top Parent universe of 18 space-time dimensions. Within that universe sub-universes may be generated through universe interactions.

3. Generate universes from vacuum states.

Once a set of universes is generated, universes may interact dynamically with universes being created or destroyed in quantum processes. It is then possible for the set of universes to achieve an equilibrium. Blaha (2024a) considered such a set of universes and found a model with a 23 universe distribution. See Figs. 6.1 and 6.2.

The physical laws within universes determine their evolution. See our previous books for details.

6.2 Cosmos Theory Brought to the Simplest Formulation

The formulation of Cosmos Theory has been reduced to the simplest terms:

1. A primordial point is the primitive term;

2. A direct generalization to a scale independent set of piecewise linear curves that combine to produce a fractal, Hilbert-like infinite curve – the postulate;

3. At each order of construction a set of dimensions (dimension array) is defined that specify a space. A space's extent is variable within a parent universe;

4. Universes may be defined as quantum field states;

5. Universes of different spaces may interact.

The procedure for creating the Cosmos parallels the creation process of Euclid's Geometry. It may be viewed as strictly mathematical – leading to the dynamics of

[23] Parts of this section and subsequent sections appeared in Blaha (2024b).

interacting universes and particles within universes. The Unified SuperStandard Theory (UST) is one of the results.[24]

Cosmos Theory is a complete Cosmology based on pure number.

6.3 Unified SuperStandard Theory

This author developed Cosmos Theory in the past 5+ years based on the prior development of the Unified SuperStandard Theory (UST) in the preceding 20 years.[25] Five years ago the author noticed a correspondence between the UST with its 256 fermion spectrum (Figs. 6.3 and 6.4), and the UST's 256 fundamental representation internal symmetry and space-time dimensions (Figs. 5.1 and 5.2), with a Cayley number 256. These features led to the author's development of Cosmos Theory.

Over the past five years the features of Cosmos Theory were developed in detail including its unification mechanism, its pattern of Cosmos Space features, and its fundamental basis.

The close, detailed, agreement between the r = 4 Cosmos Theory space structure and the UST is circumstantial evidence for Cosmos Theory. Exploration of higher dimension spaces is not possible as yet although there is suggestive data of possible external universes' influences on the structure of our universe. The Standard Model with its successes fits very beautifully within the r = 4 dimension array structure. When new fermions and their symmetries are found, then there is reason for hope that the UST will be fully vindicated, and Cosmos Theory supported.

If our universe does not have Dark Matter, a possibility, then our universe would implement a HyperCosmos space of the Second Kind.

6.4 UST Source of the Distributions of Energy, Normal Matter and Dark Matter in the Various Numbers of Particle Species

The UST has a wide range of predictions: new generations of fermions, new layers of fermions, new internal symmetries, and a Dark Matter sector. The realization of these possibilities seems somewhat unlikely at present or near term available accelerator energies.

Thus it is reasonable to examine the global possibilities due to the 256 elements in the Unified SuperStandard Theory (UST) (Cayley-Dickson number 3 Cosmos space in Fig. 3.1). These possibilities center on the relative proportions of normal and Dark matter and the distribution of energy.

We base this analysis on an equipartition of mass and energy between their various species and modes.[26]

---

[24] The UST was developed over the past 23 years by this author. It led to hypernumber quantum theories and then to Cosmos Theory.

[25] The UST is based on Two Tier quantization (2003) to eliminate perturbation theory infinities, on PseudoQuantization, which the author developed in the 1970's, and on higher derivative quantum field theories for quark confinement and modified divergence-free gravitation.

[26] Equipartition of mass and energy in Blaha (2016a) and (2018e).

6.5 Equipartition Principle for Particle Degrees of Freedom

In a closed system at equilibrium the thermal energy of a system is equally partitioned (distributed) among its degrees of freedom. This Equipartition Principle is well known. The application of this principle to the beginning of the universe *when all particles were massless* and all symmetries were unbroken suggests that the distribution of mass-energy should be the same for all degrees of freedom at that time. Thus there should be approximately equal numbers and energies of fermions and vector bosons with the same fraction of the total thermal energy.

We now estimate the relative proportion of Normal and Dark matter in the universe at its beginning based on this Equipartition Principle.

6.6 Proportion of Dark Mass-Normal Mass in the Universe

First we note that 24 of the 64 fermion species (counting quarks of each color as a separate species) in layer 1 of UST – the layer with which we are familiar are Normal matter fermions. (Our discussion is based on our Unified SuperStandard Theory.) Forty of the 64 first layer species are Dark.

The other three fermion layers are all Dark from our point of view since they have not been detected. Thus we find that 232 of the 256 fermion species are Dark yielding a *percentage of Dark Matter equal to 232/256 = 91.1%.*

Recent studies of the proportion of Dark Matter in the universe have yielded an estimate: 85% of the total mass.

Thus our rough estimate based on our fermion Equipartition Principle is approximately the same as the experimental estimate. The disparity may be due to the decay of Dark fermions to Normal fermions over time.

6.7 Proportion of Dark Mass-Energy in the Universe

We know of 12 of the 256 vector boson dimensions in the Unified SuperStandard Theory dimension array are associated with known interactions and 24 out of 256 fermions are Normal. Thus we find a total of 476 out of 512 degrees of freedom are Dark yielding a Dark mass-energy of 93% of the universes mass-energy at the beginning of the universe.

The sum of Dark energy in the universe currently has been estimated to be 68.2% of the total energy and the energy of Dark Matter is estimated to be 26.8%. The total is 95% - a value that compares favorably with our above approximate estimate of 93%. The difference in these values can be attributed to various factors. One distinct possibility is the decay of Dark mass-energy from higher layers to the known layer in the 13.8 billion years since the Big Bang.

6.8 Implications of Approximate UST Match with Experiment

The approximate match suggests the dimension array of 256 dimensions is correct as opposed to the much smaller Standard Model results. It also supports the UST proportions of Normal and Dark fermions, and the proportions of known and Dark internal symmetries—global support for the UST matching the detailed support for the

UST fermion and internal symmetry spectrums. It also buttresses the validity of Cosmos Theory.

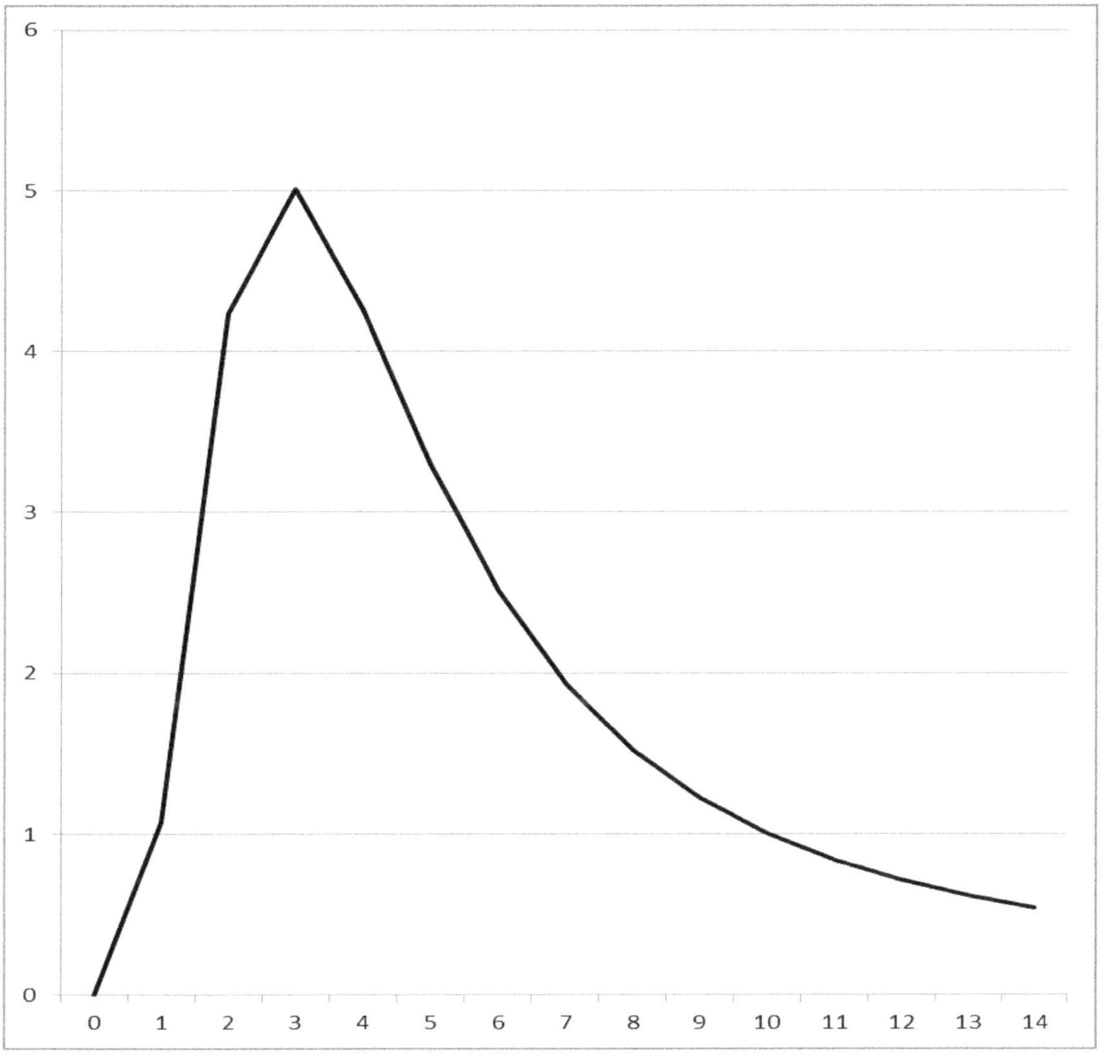

Figure 6.1. Plot of $U_g(\varepsilon(n))$ vs. the Cayley-Dickson number n. $[U_g(\varepsilon(10))] = 1$ by design. The number of universes for n > 10 is zero: $[U_g(\varepsilon(n))] = 0$ for n > 10. Small n corresponds to larger s and T_g. Large n corresponds to smaller s and T_g. The jagged curve is the result of "coarse graining" n. Fig. 11.2 shows the numbers of universes $[U_g(\varepsilon(10))]$ is 23 with one Blaha Number N = 0 parent universe and 22 "gambol" universes. See Blaha (2024a).

UNIVERSE NUMBERS

n	1	2	3	4	5	6	7	8	9	10	11
$[E_g]$	1	4	5	4	3	2	1	1	1	1	0

Figure 6.2. Table of Cayley-Dickson number n vs. the number of universes,$[U_g(\varepsilon(n))]$. Note there are no universes for HyperCosmos spaces n > 10. Thus Cosmos Theory may be limited to ten HyperCosmos spaces with this model. The universe of type n = 1 have zero space-time dimensions and 16 components in their dimension arrays. There is no n = 0 universe. See Fig. 10.1 of Blaha (2024a).

	Normal				**Dark**			
	4		**4**		**4**		**4**	

Layer 1

Generation								
1	e	q-up	ν	q-down	e	q-up	ν	q-down
2	μ	q-up	ν	q-down	μ	q-up	ν	q-down
3	τ	q-up	ν	q-down	τ	q-up	ν	q-down
4	e′	q-up	ν	q-down	e′	q-up	ν	q-down

Layer 2

Generation								
1	e	q-up	ν	q-down	e	q-up	ν	q-down
2	μ	q-up	ν	q-down	μ	q-up	ν	q-down
3	τ	q-up	ν	q-down	τ	q-up	ν	q-down
4	e′	q-up	ν	q-down	e′	q-up	ν	q-down

Layer 3

Generation								
1	e	q-up	ν	q-down	e	q-up	ν	q-down
2	μ	q-up	ν	q-down	μ	q-up	ν	q-down
3	τ	q-up	ν	q-down	τ	q-up	ν	q-down
4	e′	q-up	ν	q-down	e′	q-up	ν	q-down

Layer 4

Generation								
1	e	q-up	ν	q-down	e	q-up	ν	q-down
2	μ	q-up	ν	q-down	μ	q-up	ν	q-down
3	τ	q-up	ν	q-down	τ	q-up	ν	q-down
4	e′	q-up	ν	q-down	e′	q-up	ν	q-down

Figure 6.3. Periodic table for the Normal and Dark fermions of UST. Each fermion corresponds to a dimension. Subscripts of neutrinos and quarks are not shown. A fourth generation is indicated for each layer. There are 256 fermions corresponding to the 256 entries in the dimension array for n = 3. This figure is from Blaha (2020d) for the UST. The nesting in 4×4 blocks is evident within 8×8 blocks within a 16×16 block.

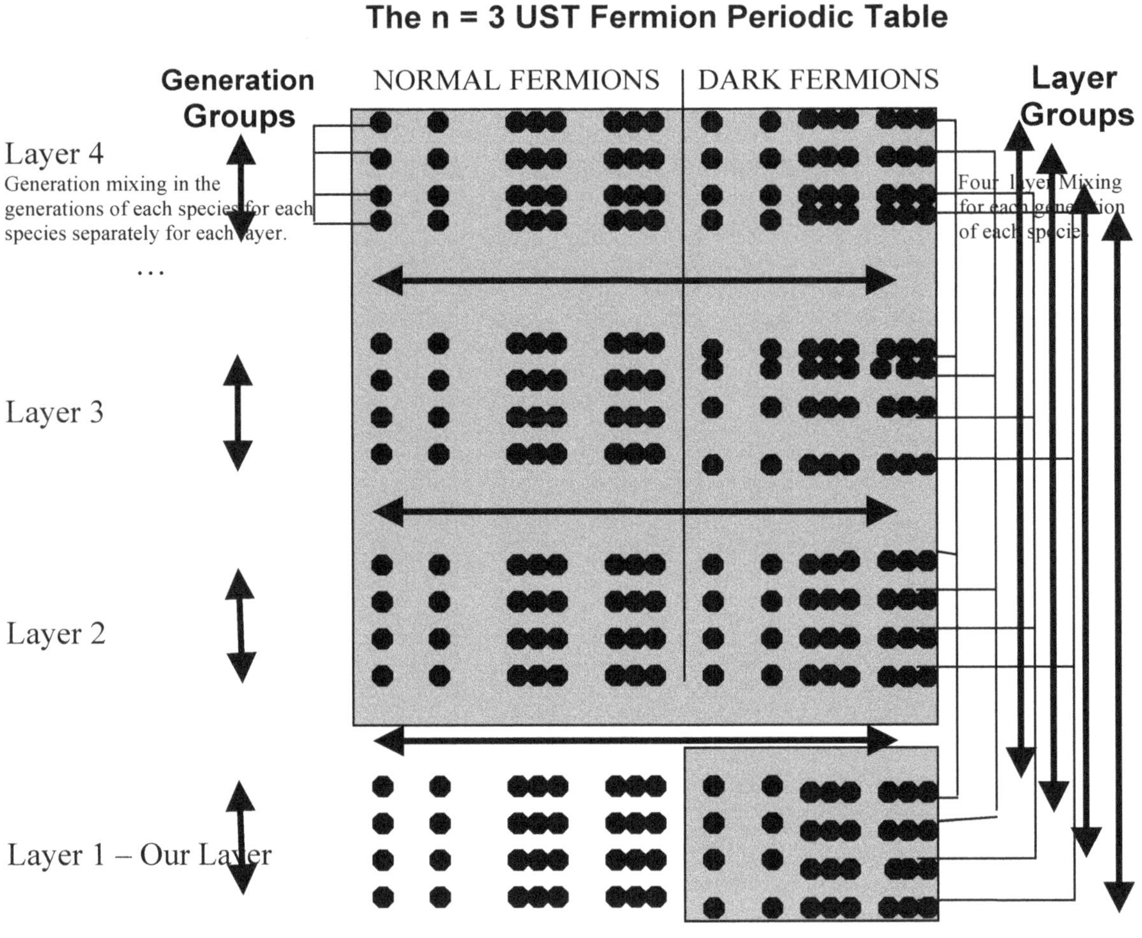

Figure 6.4. Fermion particle spectrum and partial example of pattern of mass mixing of the Generation and Layer grroups. Unshaded parts are the known fermions including an additional, as yet not found, 4th generation shown. The lines on the left side (only shown for one layer) display the Generation mixing within each layer's species. The Generation mixing applies within each layer using a separate Generation group for each layer. The lines on the right side show Layer group mixing with the mixing amongst all four layers for each of the four generations individually. There are four Layer groups. For each generation and each layer, SU(2)⊗U(1) mixes between an e-type fermion and a neutrino-type fermion. It also mixes between an up-quark-type fermion and a down-quark-type fermion. SU(3) mixes among each up-quark triplet and down-quark triplet separately. Complex Lorentz group transformations map among all four fermions: Dirac ↔ tachyon ↔ up-quark ↔ down-quark There are 256 fundamental fermions counting quarks as triplets. This figure is from Blaha (2020e) for the UST.

7. Fractal Cosmos Curve Decomposition

The Cosmos Curve is constructed from a mathematical point to a unit line segment and then to a two-dimension unit square grid. The construction from a mathematical point to a line segment relies on the sum:

$$\sum_{n=1}^{\infty} 2^{-n} = 1 \qquad (4.4)$$

The terms in this summation correspond directly to ½ of the length of columns d_c in the Cosmos Theory spectrum for negative space-time dimensions. We have called this part of the Cosmos spectrum Limos. See Fig. 3.1.

We then construct the remainder of the fractal Cosmos Curve from line segments that order by order are of length equal $\frac{1}{2}d_c$ for the HyperCosmos spaces in Fig. 3.1.

Thus the fully adjoined sequence of dimension array column lengths for $n = -\infty$ through $n = \infty$ gives a *Cosmos Curve from a zero dimension point to a two dimension filled square grid.*

The Cosmos Curve is a mathematical construct. We assume the column lengths corresponding to Cayley-Dickson numbers n define the rows and columns d_{cn} of Cosmos space dimension arrays. Cosmos space dimension arrays are not substances. They are abstract definitions of sets of dimensions. As such they conceptually correspond to the lengths of Cosmos Curve line segments.

7.1 Map from Cosmos Curve Two Dimension Grid to Cosmos Theory Spaces

We now construct a relationship between a position (x, y) within the unit square of dimension two with the corresponding Cosmos space dimension array size d_{dr}. We propose the relation:

$$d_{dr} = 2^{r+4} = 2^{8/(x+y)} \qquad (7.1)$$

where the coordinates x and y are confined to a two dimension unit grid by:

$$\begin{aligned} &0 \le x + y \le 2 \\ &0 \le x \le 1 \\ &0 \le y \le 1 \end{aligned} \qquad (7.2)$$

Eq. 7.1 implies the relation between x and y and the space-time dimension r:

$$x + y = 8/(r + 4) \qquad (7.3)$$

or

$$r = 8/(x + y) - 4 \tag{7.4}$$

We note:

If $x + y = 2 \quad r = 0$

If $x + y = 1 \quad r = 4$

If $x + y = 0 \quad r = \infty$

7.1.1 The Main Sequence of Cosmos Spaces

If we set x = y to view the diagonal, which we call *The Main Sequence of Cosmos Spaces*, we find

If $x = 1 \quad r = 0$

If $x = ½ \quad r = 4$

If $x = 0 \quad r = \infty$

giving a diagonal line for x and y for the entire HyperCosmos spectrum (where $r \geq 0$) with

$$r = 4/x - 4 \tag{7.5}$$

See Fig. 7.1. The 10 spaces of the HyperCosmos have even r values ranging from r = 0 to r = 18. The value of x corresponding to r = 18 is x = 2/11. Therefore the set of HyperCosmos spaces correspond to the x range [2/11, 1].

7.1.2 The Other Diagonal

If we set x + y = 1 then we find

If $x = 1$ and $y = 0 \quad r = 4$

If $x = y = ½ \quad r = 4$

If $x = 0$ and $y = 1 \quad r = 4$

giving the square grid's other diagonal line for x and y that has r = 4 for all x and y where x + y = 1. See Fig. 7.2. *This diagonal gives a certain prominence to r = 4, the dimension of our universe.*

7.2 Continuity of the x = y Main Sequence Diagonal in the Unit Cosmos Grid

The Main Sequence diagonal in the Cosmos Curve grid interpolates between the space-time dimensions r of HyperCosmos spaces. By eq. 7.5 we find

$$x = 4/(r + 4) \tag{7.6}$$

The HyperCosmos values of r, which is even, select a set X of corresponding x values ranging from 0 ($r = \infty$) through 1 ($r = 0$). Thus we can use this square grid diagonal to obtain a continuous line of r values for spaces intermediate between HyperCosmos spaces[27] and extending to $r = \infty$. Between the spaces in the Main Sequence of Cosmos spaces are spaces that provide transitions between Cosmos Theory spaces. These spaces might play a role if travel between universes were ever to become feasible. Chapter 12 considers "travel" between spaces via intermediate dimensions.

7.3 The Cosmos Grid

The points of the two dimension square grid, that we call the *Cosmos Grid*, generated by the fractal Cosmos Curve specify possible Cosmos spaces. We start with the y = x diagonal which we call the *Main Sequence of Cosmos spaces*. The Cosmos spaces points within the Cosmos Grid for the Limos spaces are described in section 8.3.

7.3.1 Cosmos Grid points for HyperCosmos Spaces

The Cosmos Grid points for HyperCosmos spaces are given by eqs. 7.6 and 7.5. See Fig. 7.3 for a depiction.

7.3.2 Cosmos Grid points for HyperCosmos Spaces of the Second Kind

The Cosmos Grid points for HyperCosmos spaces of the Second Kind have the values

$$x = 4/(r + 3) \tag{7.7}$$

Their dimension is

$$r = 4/x - 3 \tag{7.8}$$

The size of their associated dimension arrays is

$$d_{dr} = 2^{r+3} = 2^{4/x} \tag{7.9}$$

7.3.3 Cosmos Grid Points for Cosmos Unification Spaces

The Cosmos Grid points on the Main Sequence of Cosmos spaces of the 88 Dimension UltraUnification Space and the 42 Dimension Full HyperUnification Space (Fig. 4.6) are

$$x = 1/23 \tag{7.10}$$

and

$$x = 2/23 \tag{7.11}$$

respectively.

7.3.4 Cosmos Grid Points for Cosmos Spaces

The Cosmos Grid points corresponding to Cosmos spaces are:

---

[27] Renormalization theory studies by 't Hooft, Veltman and others have used fractiona dimensions near 4.

HyperCosmos Spaces where $x = 4/(r + 4)$

r		x
0		1
2		2/3
4		½
6		2/5
8		1/3
	…	
18		2/11
	…	
∞		0

Second Kind HyperCosmos Spaces $x = 4/(r + 3)$

r		x	
0		4/3	Off the unit grid
2		4/5	
4		4/7	
6		4/9	
8		4/11	
	…		
18		4/21	
	…		
∞		0	

Unification Spaces $x = 4/(r + 4)$

42	2/23	42 Dimension Full HyperUnification Space
88	1/23	88 Dimension UltraUnification Space

Limos spaces x points are described in section 8.3.

7.4 Correspondence Between Cosmos Curve and Internal Symmetries

The Cosmos Curve naturally leads to HyperCosmos dimension arrays, which, in turn, leads to the dynamical Internal Symmetry interactions as shown in chapter 5. We can symbolize this relation with

Cosmos Curve → Square Grid → d_{dr} → dimension array → internal symmetry interactions

Correspondingly, we found the transitions in chapter 5:

dimension array parts → internal symmetry parts → internal symmetries → interaction terms

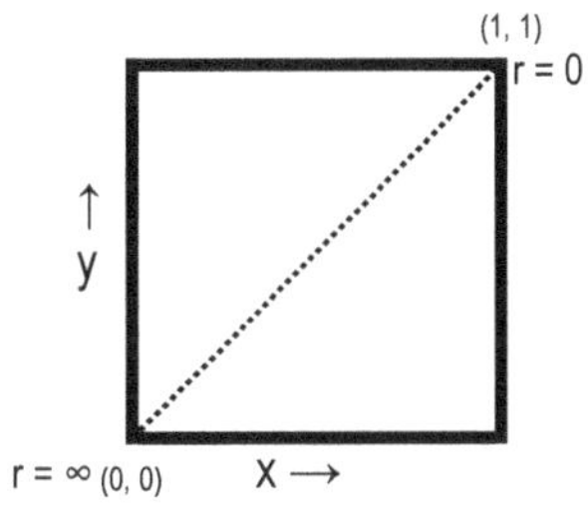

Figure 7.1. Square Cosmos Grid with a diagonal connecting r = 0 and r = ∞.

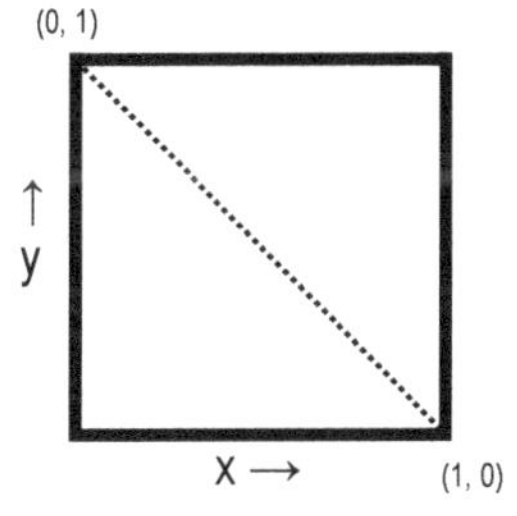

Figure 7.2. Square grid with a diagonal with each point corresponding to r = 4. These diagonals intersect at x = y = ½ where r = 4.

8. Growing Dimensions Incrementally

This chapter considers growing the dimensions of a space in a continuous manner incrementally. It begins with a discussion of the Cosmic Curve's square unit grid's derivatives. Then it considers growing a dimension from r to r + 1 in increments within the framework of the grid and also within HyperCosmos spaces.

8.1 Derivatives of Dimensions

The continuity of the grid diagonal in the x and y variables (eqs. 7.1 – 7.5) leads us to consider derivatives of d_{dn} and space-time dimension r:[28]

$$\partial d_{dn}/\partial x \quad \text{and} \quad \partial d_{dn}/\partial y \tag{8.1}$$

and

$$\partial r/\partial x \quad \text{and} \quad \partial r/\partial y$$

From eqs. 7.1 and 7.4 we find

$$\partial r/\partial x \; = \partial r/\partial y = -8/(x + y)^2 \tag{8.2}$$

$$\partial d_{dn}/\partial x \; \partial d_{dn}/\partial x \; = \partial r/\partial y \; \partial d_{dn}/\partial r \tag{8.3}$$

$$= -8/(x + y)^2 \; 2^{8/(x + y)}$$

and similarly for derivatives with respect to y. Note eqs. 8.2-8.3 imply that r increases as x + y decreases. Eq. 8.3 implies that the Cosmos' dimension array d_{dn} has a singularity at x + y = 0. All derivatives of d_{dn} with respect to x and/or y are infinite at x + y = 0.

8.2 Creation/Growth of Dimensions

As one traverses the x = y curve in the Cosmos Curve grid we find the dimension r continuously increases from zero to ∞ as x + y decreases to zero. It raises the question: How do the dimensions increase?

8.2.1 Simply Adding an Additional Dimension

One possibility is to consider the coordinate "3" dimension as growing out of a two dimensional surface. One cannot directly infinitesimally grow the third dimension from the two dimension surface. For an infinitesimal insertion makes the surface three dimensional regardless of being "infinitesimal."

---

[28] The constant n is related to r by r = 2n – 2.

8.2.2 Adding Additional Dimensions by Group Direct Products

So, in the above example, we must consider a growth path through infinitesimals that keeps the surface two-dimensional plus infinitesimal increments of a third dimension.

The spaces of the Limos sector of Cosmos Theory spaces appear to offer a solution. We note

$$d_{cn} = 2^{r/2+2} \tag{8.4}$$

where r is the dimension. Thus $r = 2\ln_2(d_{cn}) - 4$. We now define a new Limos space $s_i(r_{in})$ of dimension r_{in} where[29]

$$r_{in} = d_{cn}/4 = 2^{r/2} \tag{8.5}$$

for $r < 0$. This dimension is the r_{in} column in Fig. 8.1. The space $s_i(r_{in})$ may be taken to correspond to an orthogonal group $O(r_{in})$ of dimension r_{in} where the orthogonal group definition is extended to fractional dimensions.[30]

Then we may take a space of dimension r and increment it with s_i incrementally starting with $s_i(1/2)$ and proceeding to add to $s_i(\infty)$. The summation of the r_{in} values is one by eq. 4.4 above. The product ("summation") of the infinitesimal groups totals one unit.

For example, a sum (based on eq. 4.4) of incremental growths in dimension then may be represented as

$$SO(1, r)\otimes O(2^{-1})\otimes O(2^{-2})\otimes \ \ldots \otimes O(2^{-\infty}) \rightarrow SO(1, r + 1) \tag{8.6}$$

where O(r) represents the orthogonal group in a space of dimension r. The set of groups in eq. 8.6 is a subgroup of SO(1, r + 1) as indicated.

Thus we obtain a space with r + 1 spatial dimensions. While the space is incrementing the space effectively has a fractional dimension.

The increment process adds a sequence of infinitesimal dimensions. This process may be taken to a continuous limit. Thus the x = y grid diagonal may be realized as continuous in dimensions—based on an extension of the Limos definition of spaces for negative values of r.

---

[29] Section 3.4.2 points out "The dimension r determines the number of components in a dimension array. *The dimensions within a dimension array are degrees of freedom.* Dimension arrays define the spaces of Cosmos Theory as shown in Fig. 3.1. As pointed out in previous books this number r plays two roles: 1) the overall dimension of the space and 2) within a dimension array it specifies the number of space-time dimensions of the space." Thus the choice of dimension for a space is a matter of definition. For $r \geq 0$ we chose $r_{\text{dimension array}} = r$. We chose $r_{\text{dimension array}} = r_{in}$ for negative dimensions r.

[30] We have considered groups with fractional dimensions in past books for the case of the unitary groups.

8.3 Cosmos Grid Extension to Limos Spaces

The spaces of the Limos sector of Cosmos Theory may be mapped to the Cosmos Grid using the dimension r_{in} in eq. 8.5. With this definition of dimension we find

$$r_{in} = 8/(x + y) - 4 \tag{8.7}$$

If we assume the *Main Sequence of Cosmos spaces*: y = x then

$$r_{in} = 4/x - 4 = 2^{r/2} \tag{8.8}$$

where r is the space-time dimension column in Fig. 8.1 and

$$x = 4/(r_{in} + 4) = 4/(2^{r/2} + 4) \tag{8.9}$$

Some x, r_{in} and r values for Limos spaces

(8.10)

r	x	r_{in}
-2	8/9	½
-4	16/17	1/4
	…	
-∞	1	0

Thus the Limos spaces have x values ranging from 8/9 to 1 on the Cosmos Grid Main Sequence.

Blaha Space Number $N = o_s$	Cayley-Dickson Number n	Cayley Number d_c	Dimension Array column length d_{cn}	Dimension Array Size d_{dn}	Space-time-Dimension r	Incremental Dimension r_{in}
Limos :						
10	0	1	2	2^2	-2	½
11	-1	½	1	1	-4	¼
12	-2	¼	½	$½^2$	-6	1/8
13	-3	1/8	¼	$¼^2$	-8	1/16
14	-4	1/16	1/8	$1/8^2$	-10	1/32
		• • •				

Figure 8.1 Limos Sector of the Cosmos Theory set of spaces from Fig. 3.1. Note the additional r_{in} column.

8.4 Growth of the Cosmos Curve

We now consider growing the Cosmos Curve order by order incrementally. The orders of piecewise linear line segment lengths in the construction of a Cosmos Curve are related to the Hilbert curve construction in chapter 4:[31]

Cosmos Cayley-Dickson number $n = n_H - 1$ (4.3)
Hilbert Curve Line Length $= 2^{n_H}$
Hilbert Number of "boxes" $= 2^{2n_H}$
Cosmos Dimension Array Column Length $= 2^{n+1} = 2^{n_H}$
Cosmos Dimension Array Size (number of elements in array) $= 2^{2n+2} = 2^{2n_H}$

where n is the Cosmos Cayley-Dickson number and n_H is the order in the Hilbert curve construction.

We can create an infinitesimal growth sequence from order to order using eq. 4.4 above by infinitesimal increments:

$$2^{n+1} \rightarrow 2^{n+2}$$

where the increments are

$$2^{n+1-k}$$

for k = ∞ through 1 based on the identity

$$2^{n+1} \sum_{k=1}^{\infty} 2^{-k} = 2^{n+1} \tag{8.11}$$

The incremental growth in piecewise linear line segment lengths follows from order to order of the Cosmos Curve.

---

[31] The construction of the fractal curve corresponding to Cosmos Theory spaces was shown to be similar to the construction of the Hilbert fractal curve in Blaha (2024b).

9. Generation of Cosmos Theory Spaces from the Cosmos Grid

The HyperCosmos spaces of Cosmos Theory can be generated from the fractal Cosmos Curve. Thereby, we obtain a full specification of dimensions of the HyperCosmos Spaces starting from a zero dimension point.

Our goal was to start with the simplest possible origin—a zero dimension point that a Creator might choose and then proceed to create the Cosmos. The search for simplicity dates back to Pre-Socratic times in the West.

9.1 Dimensions of Stages in the Construction of the Cosmos Curve

We start with a zero dimension point and then use eq. 4.4 to generate a one dimension unit line segment. Then we follow a fractal construction process to create a two dimension unit square grid.

The unit square grid encompasses all the HyperCosmos spaces within it.

9.2 Extraction of Dimension Arrays of HyperCosmos Spaces

We then use the fractal grid decomposition process described in chapter 7 to extract the dimension array size d_{nr} for all HyperCosmos spaces. Thus we have the almost complete set of Cosmos spaces.

9.3 HyperCosmos Spaces of the Second Kind

The remaining spaces of Cosmos Theory, Limos spaces and HyperCosmos spaces of the Second Kind, may be generated in the unification process summarized in Fig. 4.7. See section 7.3.2 for the Cosmos Grid points of HyperCosmos spaces of the Second Kind.

9.4 Limos Sector Spaces

See section 8.3.

9.5 Cosmos Unification Spaces

See section 7.3.3 for the Cosmos Grid points of Cosmos Unification spaces.

9.4 Universes

The Cosmos Theory structure built in the above outlined manner may then be used in the construction of universes either by a creator or dynamically as discussed earlier in this book.

Thus we have generated a Cosmos from a Mathematical point.

10. Summary: Tensor Origin of Cosmos Theory

Cosmos Theory, with all its implications, is based on one mathematical fact and two assumptions. The mathematical fact follows from a consideration of the representations of the homogeneous Lorenz group in four space-time dimensions extended to other space-time dimensions r.

There is a difference between undeniable mathematical fact and *ad hoc* hypotheses. The fact in question is the number of independent anti-symmetric tensors of 0, 1, 2, … , r indices in r even space-time dimensions;

$$2^r \tag{10.1}$$

This number equals the number of elements in an irreducible γ-matrix.[32] The number of rows (columns) of γ-matrices is

$$2^{r/2} \tag{10.2}$$

This number is also twice the number of spins in r dimensions for the fermion of lowest spin in r dimensions.[33]

If we consider the form of a fermion's quantum fields,[34] the first assumption:

$$\psi_1(x) = \Sigma_{\alpha,s}[b_1(\alpha, s)u_{\alpha s}f_\alpha(x) + d^\dagger{}_1(\alpha, s)v_{\alpha s}f_\alpha{}^*(x)] \tag{10.3}$$
$$\psi_2(x) = \Sigma_{\alpha,s}[b_2(\alpha, s)u_{\alpha s}f_\alpha(x) + d^\dagger{}_2(\alpha, s)v_{\alpha s}f_\alpha{}^*(x)] \tag{10.4}$$

where α represents the Fourier momentum, then the total number of creation and annihilation operators is

$$d_{cr} = 2^{r/2 + 2} \tag{10.5}$$

where an additional factor of 2 follows from taking account of the d and $d^\dagger$ operators and a second additional factor of 2 follows due to having two quantum fields.

The column length d_{cr} specifies degrees of freedom since the associated creation and annihilation operators generate independent states.

---

[32] For our universe where r = 4 the γ-matrices are the Dirac γ-matrices.

[33] The least spin fermion quantum field in a space-time of dimension r has the spin $s = ½\,(2^{r/2 - 1} - 1)$

[34] In our PseudoQuantum formulation of fermions in the 1970's and recvently we define two wave functions for each fermion for important reasoms presented in my earlier papers and books.

10.1 Dimension Arrays

We may define an array, the dimension array, by introducing a group index for a group with the same number of components as d_{cr} and use it to define a dimension array with d_{dr} elements:

$$d_{dr} = d_{cr}^{\ 2} = 2^{r+4} \tag{10.6}$$

The quantum fields with index a become:

$$\psi_{1a}(x) = \Sigma_{\alpha,s}[b_{1a}(\alpha, s)u_{\alpha s}f_\alpha(x) + d^\dagger{}_{1a}(\alpha, s)v_{\alpha s}f_\alpha{}^*(x)]$$
$$\psi_{2a}\,(x) = \Sigma_{\alpha,s}[b_{2a}(\alpha, s)u_{\alpha s}f_\alpha(x) + d^\dagger{}_{2a}\,(\alpha, s)v_{\alpha s}f_\alpha{}^*(x)]$$

The number of additional components is set equal to the number of operators, eq. 10.4, to give the group the same number of degrees of freedom. As a result d_{dr} may be represented in the form of a square array. We call it the *dimension array*. We define a space for each space-time dimension r with a corresponding dimension array. *We relate space dimensions to the set of creation/annihilation operators (with internal symmetries) since they both specify degrees of freedom.*

We make the second assumption:

We define spaces, which each have an associated dimension array that specifies a set of dimensions, and that specify the spectrum of each set of fundamental fermions, of scalar particles, and of symmetries (including space-time dimensions r and internal symmetry dimensions.)

Each space-time has an even number of dimensions. Odd space-time dimension cases are ruled out because they would have space-time dimension arrays that are redundant with the even space-time dimensions' dimension arrays. The numbers of their spins in odd dimensions are the same as the next lower even dimensions leading to redundant dimension arrays.

The dimension arrays and their associated spaces result.[35] We have called the theory of these spaces Cosmos Theory. Cosmos Theory is described in detail in our previous books.

The set of positive space-time dimension spaces ranges from r = 0 to r = ∞. We arbitrarily set the number of spaces with universes to ten. We may base this choice on a Planckian model for universes presented in Blaha (2024b).

Below we show that the column lengths of dimension arrays form a sequence that mirrors the piecewise linear elements of the fractal Hilbert curve leading us to characterize the HyperCosmos spaces as forming part of the *fractal Cosmiic Curve*.

[35] The appearance of powers of 2 is ultimately the result of the number of independent tensors in each space-time.

10.2 Extension of Spaces to Fractional Space-time Dimensions

The space-time dimensions that we have considered above are non-negative, even integer-valued dimensions greater than or equal to zero. We may introduce more dimensions by descending to negative dimensions as shown in Fig. 10.1. Spaces with negative integer dimensions are called the Limos spaces. They are used in gambol theory in Blaha (2024a).

We introduce positive (and now negative) Cayley-Dickson numbers n related to the positive (and negative) space-time dimensions r by

$$r = 2n - 2 \tag{10.7}$$

with the result

$$d_{cn} \equiv d_{cr} = 2^{n+1} = 2^{r/2+2} \tag{10.8}$$

$$d_{dn} \equiv d_{cr} = 2^{2n+2} = 2^{r+4} \tag{10.9}$$

We will use n to relate the dimension arrays to a Hilbert-like fractal curve.

Our ProtoCosmos model created a set of spaces (Fig. 10.1) with negative dimensions and fractional dimension arrays. We used them in our gambol studies (Blaha (2024a)) and to extend the Cosmos Curve to a zero dimension point. See below and Blaha (2024b).

10.3 The Fractal Cosmos Curve

We now use the fractal relations first found in Blaha (2024b). The fractal construction of the Cosmic Curve is based on the association of dimension array column lengths[36] and the orders of piecewise linear line segment lengths that are used to form a fractal curve. Eq. 10.10 relates the *Cosmos Curve* construction to the Hilbert curve construction.[37]

Cosmos Cayley-Dickson number $n = n_H - 1$ (10.10)
Hilbert Curve Line Length $= 2^{n_H}$
Hilbert Number of "boxes" $= 2^{2n_H}$
Cosmos Dimension Array Column Length $= 2^{n+1} = 2^{n_H}$
Cosmos Dimension Array Size (number of elements in array) $= 2^{2n+2} = 2^{2n_H}$

where n is the Cosmos Cayley-Dickson number and n_H is the order in the Hilbert curve construction. Figs. 10.2 – 10.4 show the map between Cosmos Curve diagrams and equivalent Hilbert curve diagrams.

The non-negative n (or r) dimension array column lengths combine to generate a two dimension filled square fractal grid from a one dimension line segment.

---

[36] Order by order (dimension by dimension) the dimension array column lengths map to the length of the corresponding Hilbert curve piecewise line lengths. Blaha (2024b).

[37] The construction of the fractal curve corresponding to Cosmos Theory spaces was shown to be similar to the construction of the Hilbert feactal curve in Blaha (2024b).

The negative n (r) dimension array column lengths for n = -2 through n = -∞ combine to form a one dimension line element of length 1 equal to the array column length 1 for n = -1 from a zero dimension point due to the identity

$$\sum_{n=1}^{\infty} 2^{-n} = 1 \qquad (10.11)$$

If we fully adjoin the dimension array column lengths for n = –∞ through n = ∞ then we have the *Cosmic Curve from a zero dimension point to a two dimension filled square grid.*

The fractal dimension generation is paralleled by the generation of dimension arrays at each step in its construction.

10.4 The Nesting of Cosmos Dimension Arrays

Figs. 10.2, 10.3, 10.7, 10.8, and 10.9 show the nesting of dimension arrays and Hilbert construction diagrams. For each n (and Hilbert order) the nth array (and diagram) contains four copies of the $(n-1)^{th}$ array (and diagram).

The nesting property carries through for the fermion spectrum and group symmetry arrays. Figs .10.4 and 10.5 show the nested fermion structure for our n = 3 universe. Figs .10.6 and 10.7 show the nested symmetry group structure for our n = 3 universe. Figs. 10.8 and 10.9 show the nested dimension array structure for the n = 4 Megaverse.

The nesting of the group symmetries is modified after the quartering (or quadrupling) to connect various quarters together. Some of the symmetry groups are defined to connect quarters together using Connection Groups that we have described in earlier books such as Blaha (2023d).

If one considers the known Standard Model part within these diagrams one sees it fits in the nested dimension arrays like the pieces in a puzzle. Thus there is experimental support for Cosmos Theory in The Standard Model.

10.5 Euclidean View of the Creation of Cosmos Theory

With the basis of Cosmos Theory in the tensor structure of spaces we now have a complete theory with fundamental assumptions and a construction process modeled on the fractal Cosmic Curve that parallels Euclid's formulation of Geometry. The theory is purely mathematical: a mirror of Reality.

10.6 Analogous γ-Matrix Features

The even space-time dimension γ-matrices exhibit properties similar to the Cosmos Theory dimension arrays:

1. Dimension arrays have a size that is 16 times the size of γ-matrices.

2. The γ-matrices are square. They are r matrices of size $2^{r/2}$ by $2^{r/2}$. As the even space-time dimension r increases by 2 the γ-matrices row and column sizes

double. The γ-matrices quadruple as a result. For example the r = 6 γ-matrices are 8 by 8 matrices containing a quadruple of r = 4 γ-matrix parts.

3. The γ-matrices exhibit a form of nesting in quadruples. Fig. 10.10 displays nesting from r =2 (Pauli matrices) to r = 4 and then to r =6 quadrupling at each dimension increase by 2.

4. Fractional γ-matrices of negative dimension n may be defined by quartering the γ-matrices of the next higher space-time dimension. These fractional matrices may be of interest for gambol quantum field theory. This topic has been considered in our previous books.

The Hilbert curve features, Cosmos dimension array features, and γ-matrix structural features are the same.

10.7 Second Kind HyperCosmos Spaces

The Second Kind HyperCosmos spaces are generated with dimension arrays that are half the size of HyperCosmos dimension arrays. They may be interpreted as the same as HyperCosmos dimension arrays but without a Dark sector. Their n^{th} space dimension array size d_{dN2} is (Fig. 10.1)

$$d_{d2n} = 2^{r+3} \tag{10.12}$$

See Blaha (2024a) and earlier books for details.

10.8 The Cosmic Curve and Unification and Fundamental Reference Frame (FRF)

In Blaha (2023a) we developed a unification of the ten HyperCosmos spaces and the 10 Second Kind HyperCosmos in a 42 space-time dimension Full HyperUnification space. See Fig. 10.11. The purpose was to create a space from these 20 spaces that supported a combined General Relativistic and symmetry group transformation that would generate the dimension arrays of all 20 spaces from one dimension located in a Fundamental Reference Frame (FRF).

In section 3.3.1 Of Blaha (2023a) we described the nature of an FRF. A dimension array Reference Frame treats a dimension array with d_{dr} components as a vector. The vector is subject to a combined General Relativistic – Internal Symmetry transformation to produce a new dimension array in vector form. One may chose the Reference Frame to be an FRF where the vector has only one non-zero dimension. This FRF is analogous to the rest frame of a particle with only one non-zero momentum component: its energy equals its mass. The analogy to Special Relativistic transformations from a rest frame to a moving frame is clear.

Using FRF's allows the dimension array of any space to be reduced to one non-zero component.

An examination of Fig. 10.11 shows that the 10 dimension arrays' column lengths and 10 Second Kind HyperCosmos dimension arrays' column lengths combine to form the 42 space-time dimension Full HyperUnification space shown in Fig. 10.12. The column lengths of the HyperCosmos part are a subset of the column lengths in the Cosmic Curve. See Blaha (2023a) for details.

The dimension array of the 42 space-time dimension Full HyperUnification space can be treated as a vector in an 88 space-time dimension UltraUnification (UU) space. Then an FRF may be defined for it with only one non-zero dimension that enables the UU dimension array to be generated from the FRF with only one non-zero dimension.

Thus the UU gives a reduction of all Cosmos Theory dimension arrays to one primordial non-zero dimension – an ultimate form of unification of all Cosmos Theory spaces. See Fig. 10.12 for the structure of the complete set of Cosmos spaces. There are four levels in the Cosmos, which is the most economical approach to generating the Cosmos. The one primordial dimension in UU can be fractionated into an infinite dust of fractionated dimensions using Limos.

The spaces of Cosmos Theory (Fig. 10.12) *are*

!0 HyperCosmos spaces + their 10 HyperUnification spaces + (10.13)
+ 10 Second Kind HyperCosmos spaces +
+ the 10 Second Kind HyperUnification spaces +
+ the 42 Dimension Full HyperUnification Space +
+ the 88 Dimension UU space = 42 spaces
+ Limos[38](not shown in Fig. 10.12)

10.9 The Two Dimension Square Grid

The Cosmic Curve generates a two dimension filled square grid. This grid might be viewed as the beginning of a new Cosmos spaces spectrum. It embodies the 4 dimensions of the 2×2 n = 0 dimension array d_{d0}, which is the precursor of the ten HyperCosmos spaces. It might be viewed as a continuous grid corresponding to d_{d0}.

The four corners of the Cosmos Curve square grid is mapped to the four dimensions of an n = 0 dimension array. We can associate that dimension array with the $n_H = 1$ Hilbert order diagram and proceed to construct a new Hilbert-like curve and a new associated sets of Cosmos dimension arrays from it. At each order of construction we take the boxes of type C in Fig. 10.2 and transform them to boxes of type D filling each with a "color pattern." As the order n_H increases, an irregularly shaped figure is generated that becomes a set of interconnected islands of "Cosmic" color.

This new construction of interconnected islands might be called a *SuperCosmos* archipelago. It corresponds order by order to supersets of Cosmos spaces with filled grids used to introduce continuity between dimensions in each 4 by 4 block. Thus a continuum is established between the elements of each diagram. The elements are, of course, Cosmos Theory sets of spaces.

[3838] Limos spaces with their negative space-time dimensions are not part of the Full HyperUnification space or the UltraUnification space because they do not participate in the unification of the twenty HyperCosmos spaces.

10.10 The Language of the Cosmic Curve

The Hilbert curve and thus the Cosmic Curve may be viewed as generated according to a language ((L System). The language specification is

Alphabet: α, β
Constants: φ + −
Axiom: α

Production Rules:
α → +βφ−αφα − φβ+
β→ −αφ+βφβ+φα-

The language interpretation of the Hilbert and Cosmic Curves introduces an interesting new view of Cosmos Theory raising the issue of a Cosmic Language. Languages have been associated with Physics previously. See Blaha (1998) and (2005c) for detailed examples.

10.11 Comparison between Fractal Formulation and Creation/Annihilation Operator Formulation

The fractal formulation, and the creation/annihilation operator based formulation, offer somewhat different approaches to Cosmos Theory. The fractal formulation, which has a Hilbert-like curve view: the Cosmic Curve, offers the perhaps a simpler, more elegant formulation. The fractal formulation has the same nesting as the Cosmos Theory dimension arrays.

The fractal formulation does not have some significant features of the Creation/Annihilation Operator (CAO) formulation :

1. ***The basis of the CAO formulation in the number of independent tensors singles out the CAO formulation as fundamental.***

2. The choice of even number space-time dimensions only is natural in the CAO formulation since the total number of creation and annihilation operators in an odd dimension space equals the number in the next lower even dimension – thus giving a potential redundancy that is removed by choosing even dimensions only.

3. The CAO formulation creates a dimension array that has extended General Relativistic – Internal Symmetry transformations that parallel the corresponding transformations of the creation/annihilation operators of fermion fields.in General Relativity.

4. The CAO formulation allows one to define a Fundamental Reference Frame (FRF) that can map a single primordial dimension to complete dimension arrays, which introduces a fundamental simplicity in the Cosmos.

5. Space-time dimensions may be extracted from a dimension array using eq. 10.2.

6. Manipulation of the quantum fields based on the CAO formalism enables the creation of totally unified quantum field theories through the generation of 42 and 88 space-time dimension spaces.

We conclude the CAO tensor-based formalism is the correct basis for Cosmos Theory and the UST for our universe.

10.12 Future Directions

The fundamental fact of Cosmos Theory, the number of independent tensors in a space, and the assumptions that then lead directly to the Cosmos spaces spectrum, make Cosmos Theory the most likely theory of the Cosmos.

This conclusion motivates the following program for Elementary Particles:

1. Study other additional features of Cosmos Theory.

2. Perform experiments to expand the fermion spectrum to a fourth generation and to three additional layers.

3. Explore the implications of the Generation and Layer groups – particularly for the expansion of the Cabibbo-Kobayashi-Maskawa (CKM) matrix.

4. Look for Astrophysical support for Cosmos Theory – particularly for the Megaverse and higher dimension universes.

The spectrum of particles and symmetries needs to be expanded from The Standard Model to the predictions of Cosmos Theory and the Unified SuperStandard Theory (UST).

Success in these endeavors will lead to a new Physics. The possibility of several new accelerators operating at 3 to 8 times current maximum energies is encouraging.

This author *solely* developed Cosmos Theory over the past five years beginning with The Standard Model, and with his previous work.

He followed the path:

The Standard Model →

Unified SuperStandard Theory →

Hypercomplex number based extensions →

Cosmos Theory →

Fractal Cosmos Curve Theory →

Tensor based Fractal Cosmos Theory

COSMOS SPACES SPECTRUM

Blaha Space Number $N = o_s$	Cayley-Dickson Number n	Cayley Number d_c	Dimension Array column length d_{cn}	Dimension Array Size d_{dn}	Space-time-Dimension r
0	10	1024	2048	2048^2	18
1	9	512	1024	1024^2	16
2	8	256	512	512^2	14
3	7	128	256	256^2	12
4	6	64	128	128^2	10
5	5	32	64	64^2	8
6	4	16	32	32^2	6
7	**3**	**8**	**16**	$\mathbf{16^2}$	**4**
8	2	4	8	8^2	2
9	1	2	4	4^2	0
10	0	1	2	2^2	-2
Limos :					
11	-1	½	1	1	-4
12	-2	¼	½	$½^2$	-6
13	-3	1/8	¼	$¼^2$	-8
14	-4	1/16	1/8	$1/8^2$	-10

•
•
•

HYPERCOSMOS OF THE SECOND KIND SPACES SPECTRUM

Blaha Space Number $N = O_s$	Cayley-Dickson Number n	Cayley Number d_c	Dimension Array size d_{dN2}	Space-time-Dimension r	CASe Group $su(2^{r/2},2^{r/2})$ CASe
0	10	1024	1024×2048	18	su(512,512)
1	9	512	512×1024	16	su(256,256)
2	8	256	256×512	14	su(128,128)
3	7	128	128×256	12	su(64,64)
4	6	64	64×128	10	su(32,32)
5	5	32	32×64	8	su(16,16)
6	4	16	16×32	6	su(8,8)
7	**3**	**8**	$\mathbf{8 \times 16}$	**4**	**su(4,4)**
8	**2**	4	4×8	2	su(2,2)
9	**1**	2	2×4	0	su(1,1)
10	0	1	1×2	-2	
11	-2	½	½	-4	

Figure 10.1. The HyperCosmos, Limos, and the HyperCosmos of the Second Kind, space spectrums.

n_H = 1 Hilbert Line

A

Extend to make a Cosmos Cayley-Dickson n = 0 Dimension Array Box (of 4 dimensions)

B

Hilbert Order n_H = 2 Fractal View, a Cayley-Dickson n = 1 Dimension Array Box (of 16 dimensions)

C

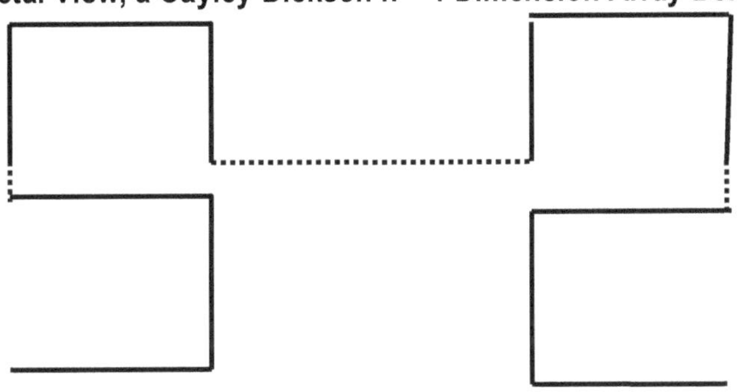

Hilbert Order n_H = 2 and n = 1 Cosmos View

D

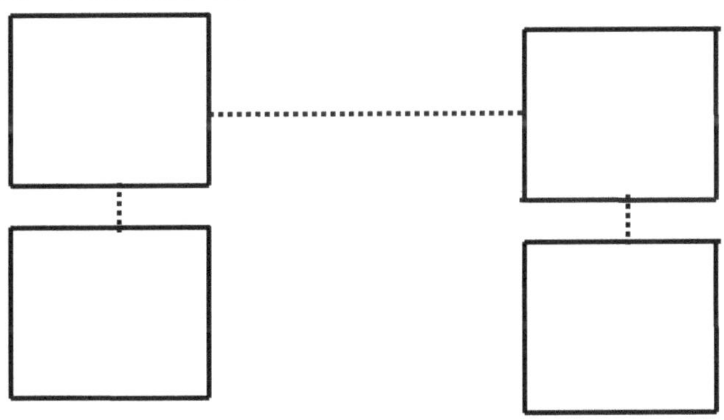

Cosmos Cayley-Dickson n = 1 Dimension Array Form (of 16 dimensions)

E

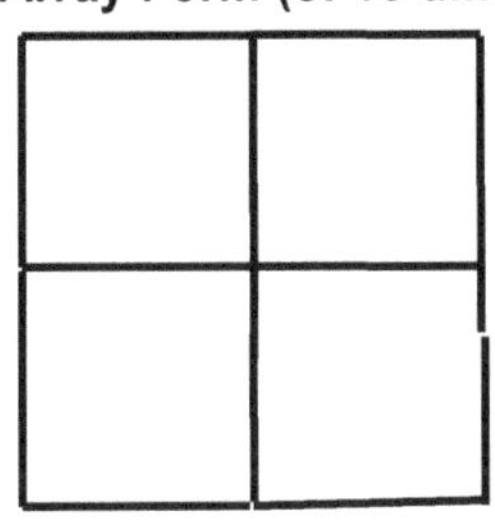

Figure 10.2. The transition from the Hilbert curve Fractal diagrams to the Cosmos Theory dimension array diagrams. Every Hilbert fractal diagram has an equivalent Cosmos Theory dimension array.

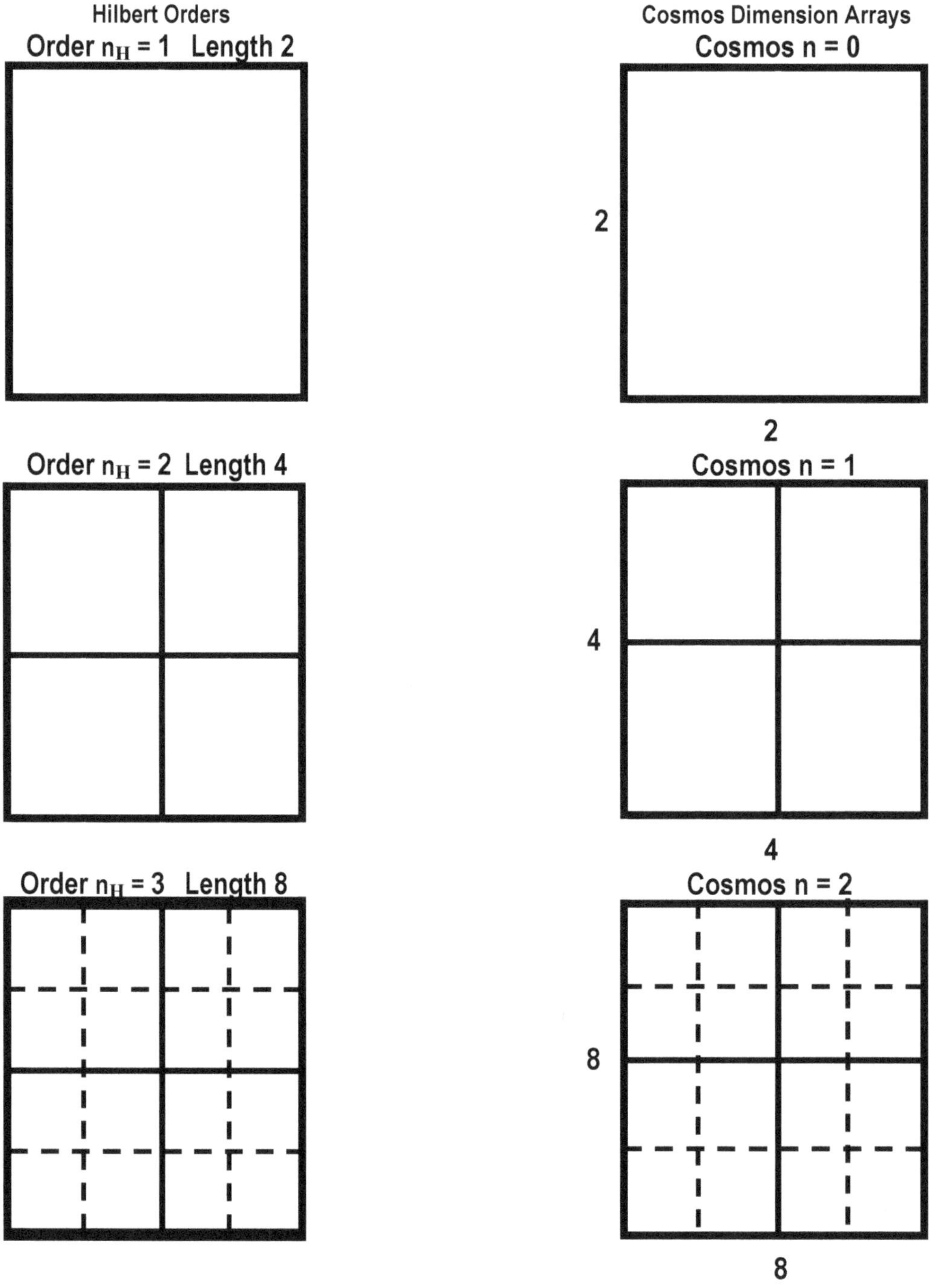

...

Figure 10.3. Low order Hilbert line segments depicted as nested sets of boxes as in Fig. 10.2 compared to corresponding Cosmos Theory dimension arrays ordered by Cayley-Dickson numbers. Dimension array sizes are displayed.

	Normal 4		Normal 4		Dark 4		Dark 4	

Layer 1

Generation								
1	e	q-up	v	q-down	e	q-up	v	q-down
2	μ	q-up	v	q-down	μ	q-up	v	q-down
3	τ	q-up	v	q-down	τ	q-up	v	q-down
4	e′	q-up	v	q-down	e′	q-up	v	q-down

Layer 2

Generation								
1	e	q-up	v	q-down	e	q-up	v	q-down
2	μ	q-up	v	q-down	μ	q-up	v	q-down
3	τ	q-up	v	q-down	τ	q-up	v	q-down
4	e′	q-up	v	q-down	e′	q-up	v	q-down

Layer 3

Generation								
1	e	q-up	v	q-down	e	q-up	v	q-down
2	μ	q-up	v	q-down	μ	q-up	v	q-down
3	τ	q-up	v	q-down	τ	q-up	v	q-down
4	e′	q-up	v	q-down	e′	q-up	v	q-down

Layer 4

Generation								
1	e	q-up	v	q-down	e	q-up	v	q-down
2	μ	q-up	v	q-down	μ	q-up	v	q-down
3	τ	q-up	v	q-down	τ	q-up	v	q-down
4	e′	q-up	v	q-down	e′	q-up	v	q-down

Figure 10.4. Periodic table for the Normal and Dark fermions of UST. Each fermion corresponds to a dimension. Subscripts of neutrinos and quarks are not shown. A fourth generation is indicated for each layer. There are 256 fermions corresponding to the 256 entries in the dimension array for n = 3. This figure is from Blaha (2020d) for the UST. The nesting in 4×4 blocks is evident within 8×8 blocks within a 16×16 block.

The n = 3 UST Fermion Periodic Table

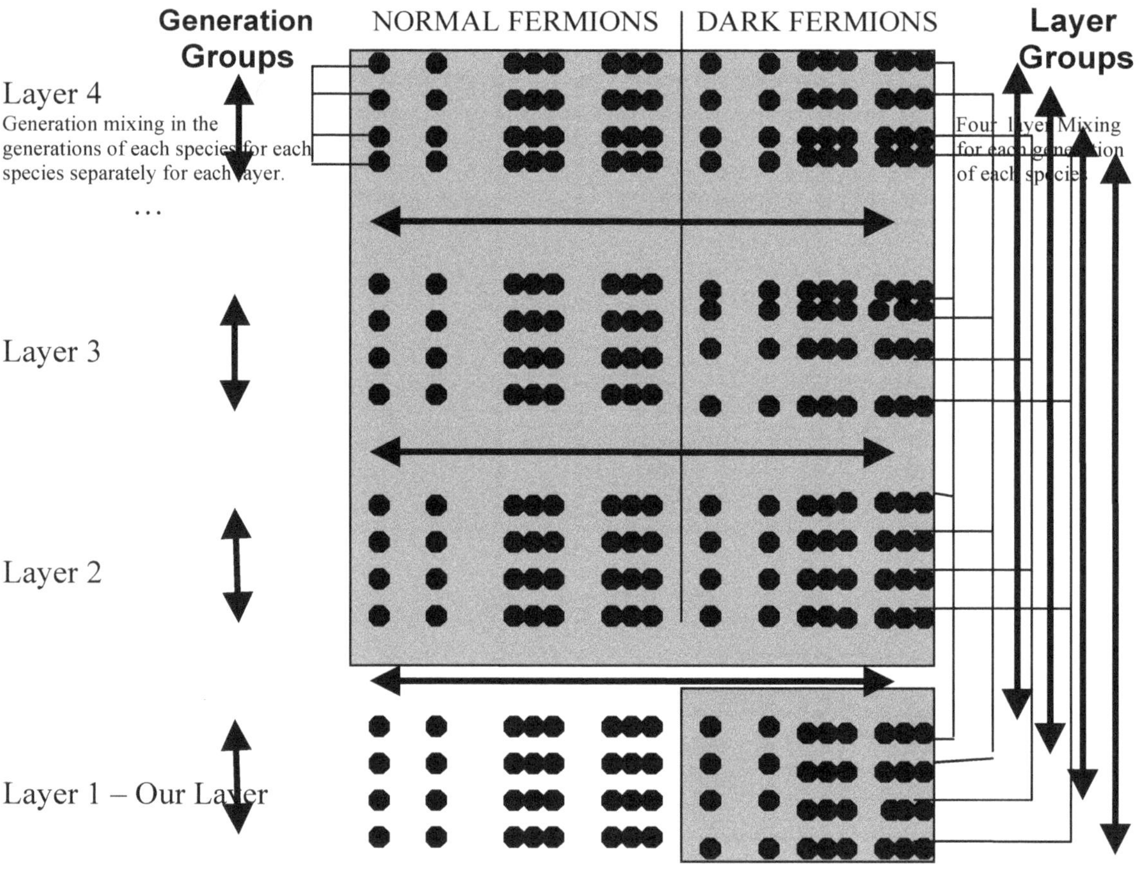

Figure 10.5. Fermion particle spectrum and partial example of pattern of mass mixing of the Generation and Layer grroups. Unshaded parts are the known fermions including an additional, as yet not found, 4th generation shown. The lines on the left side (only shown for one layer) display the Generation mixing within each layer's species. The Generation mixing applies within each layer using a separate Generation group for each layer. The lines on the right side show Layer group mixing with the mixing amongst all four layers for each of the four generations individually. There are four Layer groups. For each generation and each layer, SU(2)⊗U(1) mixes between an e-type fermion and a neutrino-type fermion. It also mixes between an up-quark-type fermion and a down-quark-type fermion. SU(3) mixes among each up-quark triplet and down-quark triplet separately. Complex Lorentz group transformations map among all four fermions: Dirac ↔ tachyon ↔ up-quark ↔ down-quark There are 256 fundamental fermions counting quarks as triplets. This figure is from Blaha (2020e) for the UST.

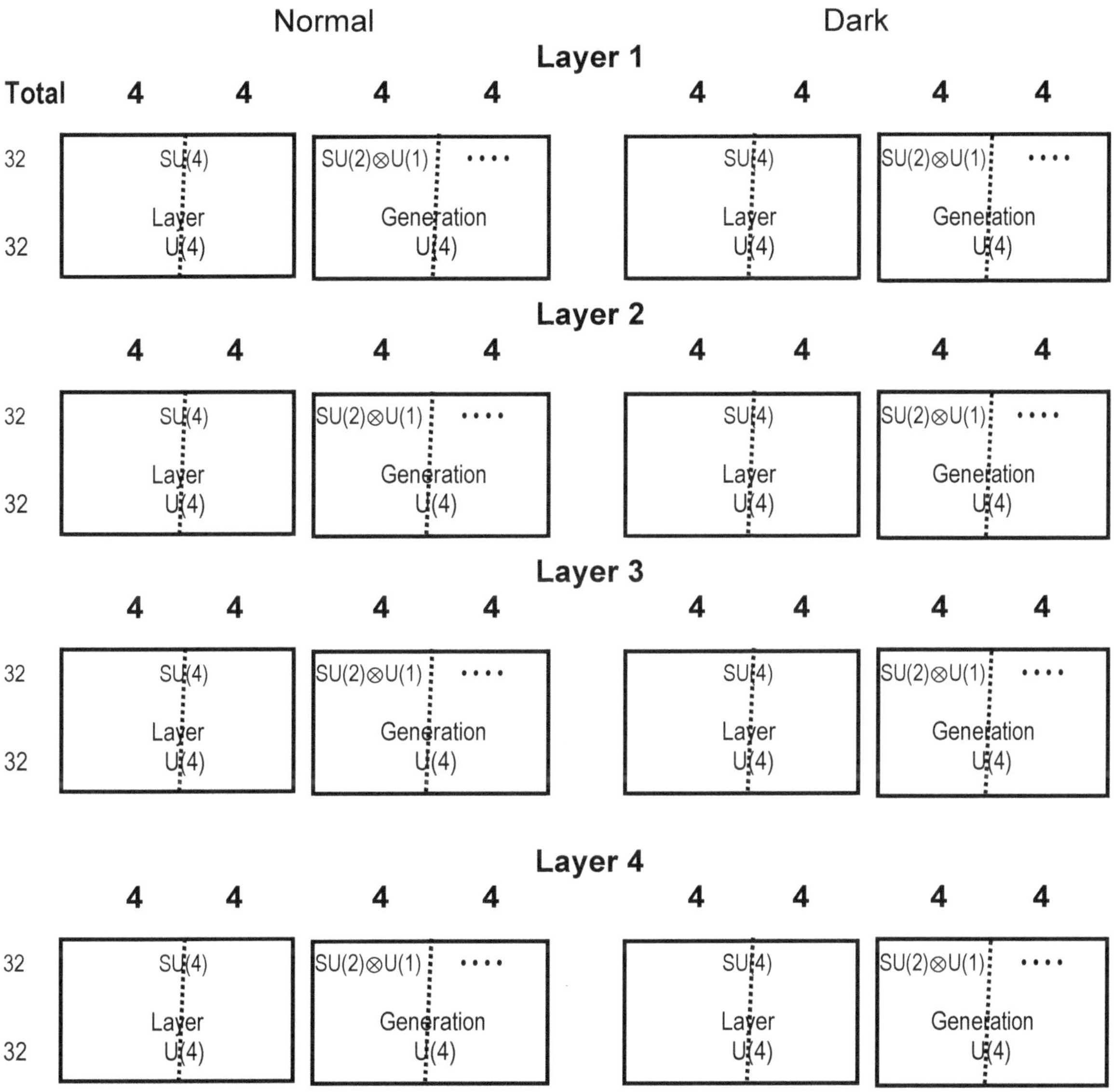

Figure 10.6. Schematic for the Normal and Dark symmetries of UST. The real and imaginary dimensions are counted separately. For example SU(4) has four real dimensions and four imaginary dimensions. Each layer has 32 + 32 = 64 dimensions. There are 256 dimensions corresponding to the 256 entries in the dimension array for UST where n = 3. The 32 • 's form sets of 4 dimensions were allocated initially to 4D space-times. These dots are mapped to SL(2, **C**) and 7 SU(2)⊗U(1) Connection groups in Fig. 10.7. Three groups unite layers, and 4 groups unite corresponding Normal and Dark layers This figure is from Blaha (2020d) for the UST.

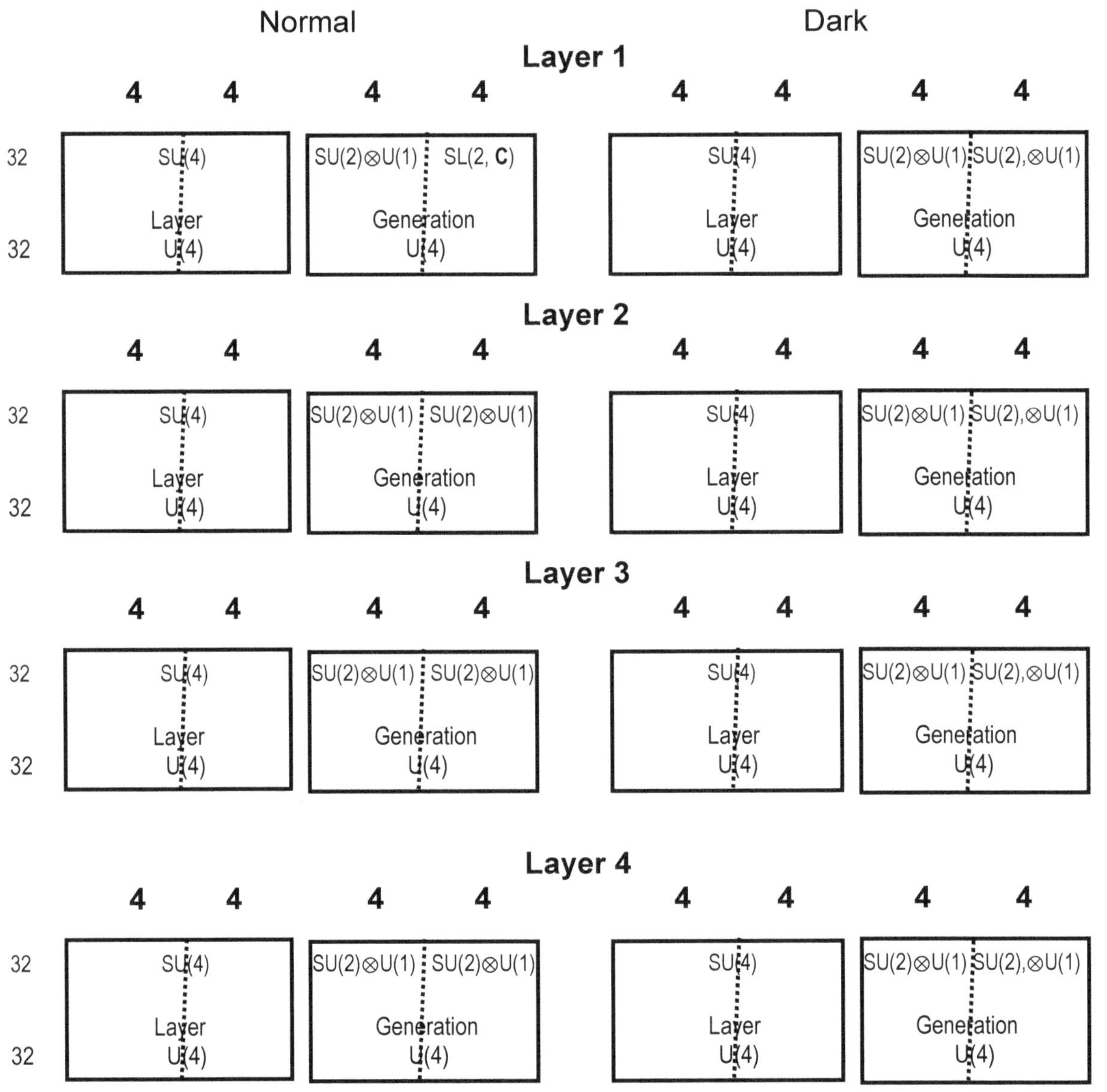

Figure 10.7. Schematic for the Normal and Dark symmetries of UST. Seven sets of dots in Fig. 10.6 are replaced by SU(2)⊗U(1) Connection groups. SL(2, C) represents the Lorentz group $SO^{+}(1,3)$. This figure is from Blaha (2020d) for the UST.

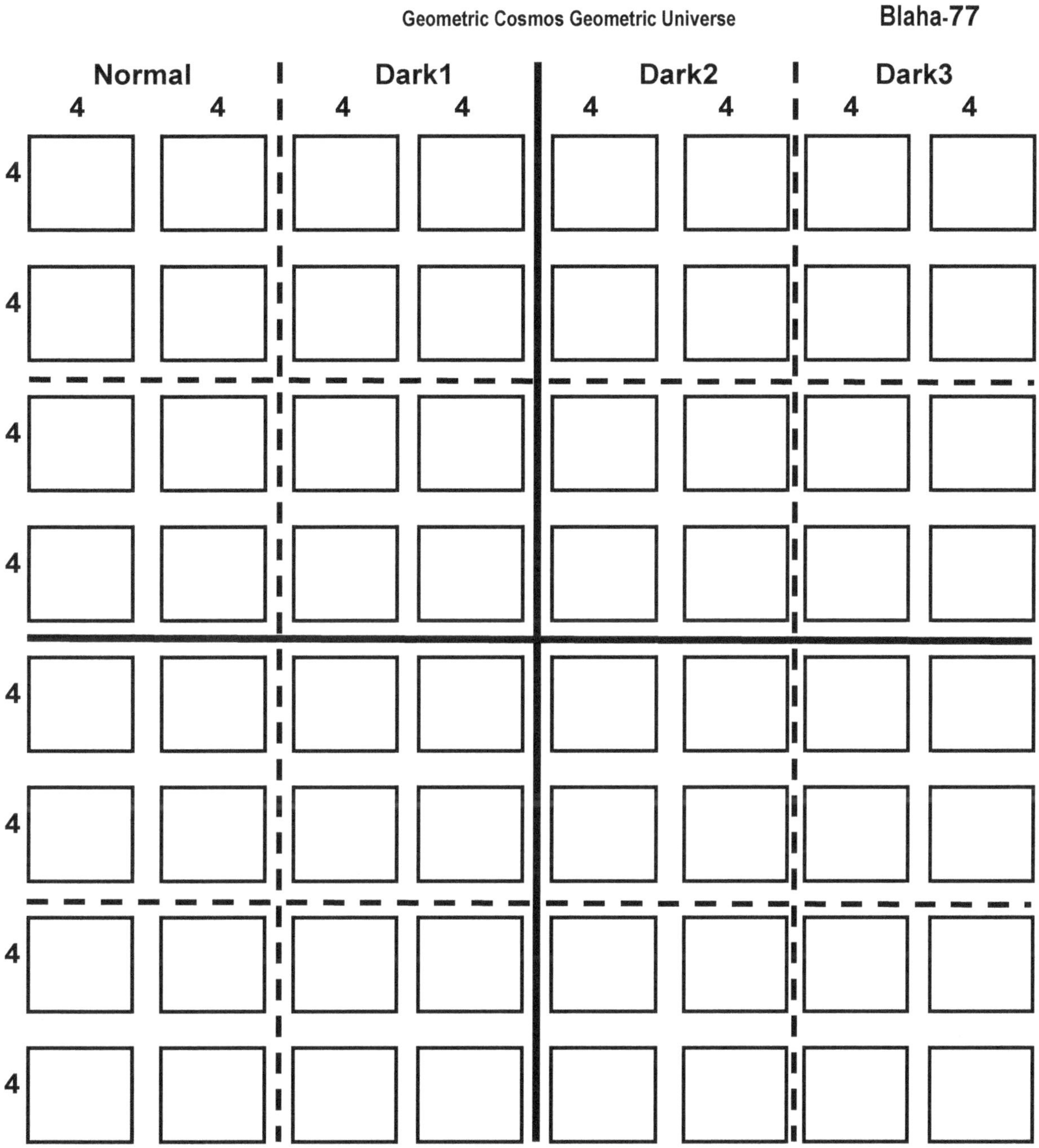

Figure 10.8. The n = 4 Megaverse dimension array. Four quadruplings are visible. Each quadrant is an n = 3 UST dimension array. There are sixty-four 4×4 n = 1 blocks. Each 8×8 quadrant is an n = 2 dimension array. Each 16×16 quadrant is an n = 3 dimension array. An exact match with Hilbert orders nesting is evident. This figure is from Blaha (2020d) for the UST.

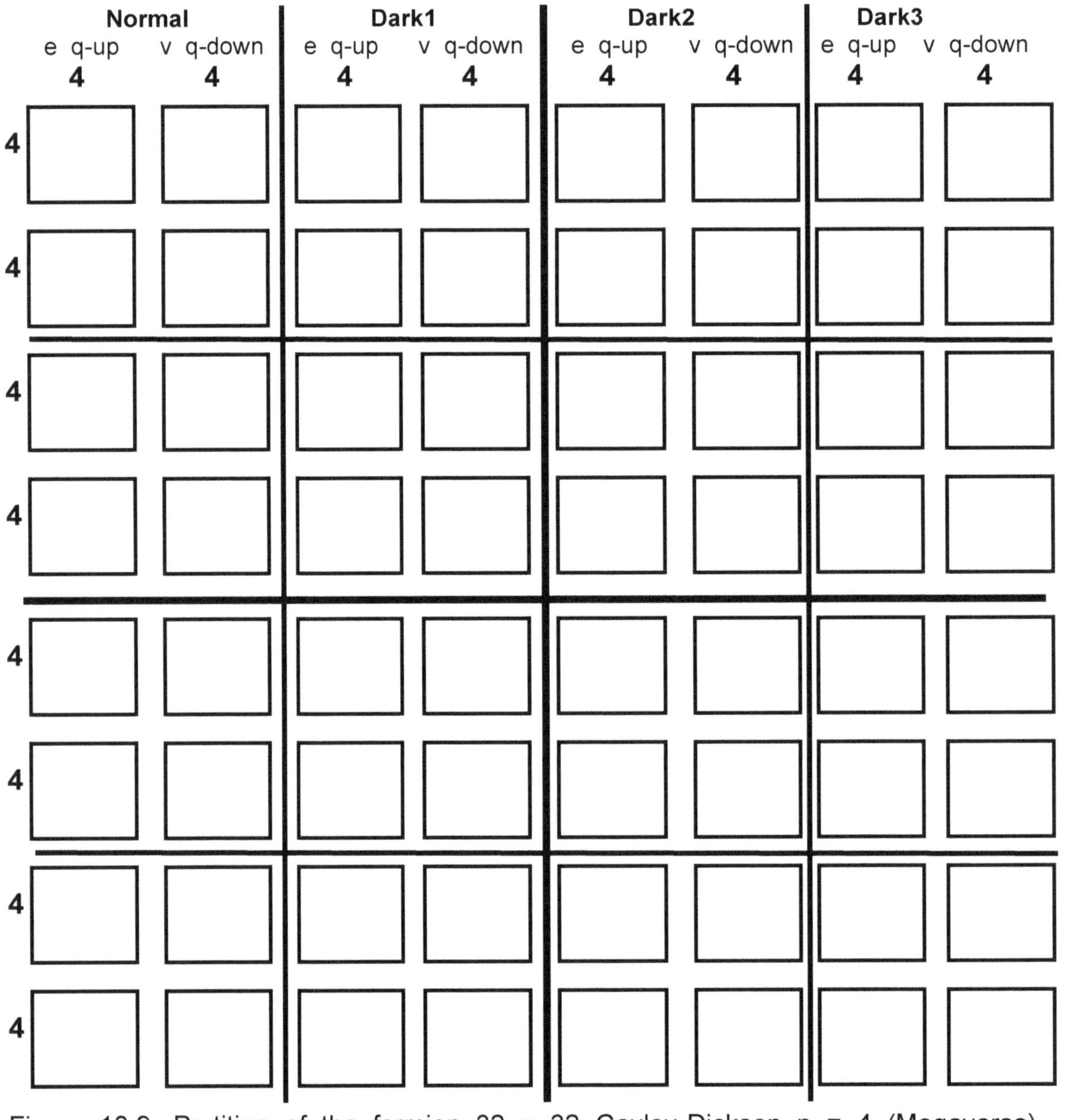

Figure 10.9. Partition of the fermion 32 × 32 Cayley-Dickson n = 4 (Megaverse) dimension array. Four quadruples are visible. There are sixty-four 4 × 4 n = 1 blocks. Each 8 × 8 quadrant is an n = 2 dimension array. Each 16 × 16 quadrant is an n = 3 dimension array. An exact match with Hilbert orders is evident. The label e q-up indicates a charged lepton – up-type quark pair. The label v q-down indicates a neutral lepton – down-type quark pair, and so on. This figure is from Blaha (2020i) for the UST and Megaverse.

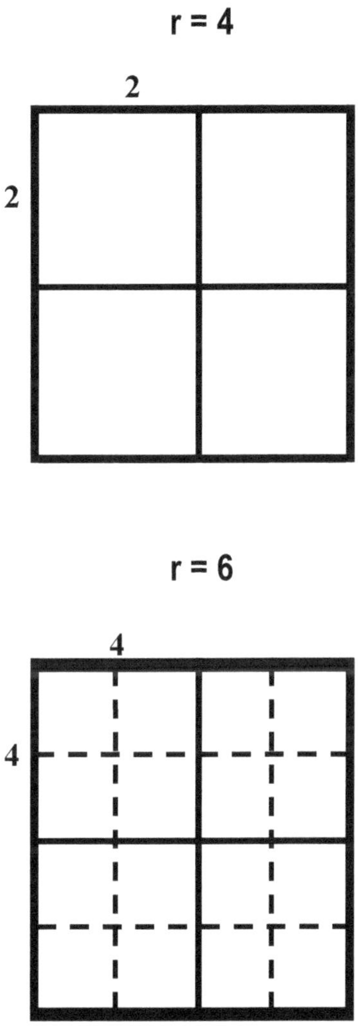

Figure 10.10. The γ-matrices for r = 4 and r = 6 space-time dimensions. The matrices exhibit a nested structure quadrupling as the dimension increases by 2. For example, the r = 4 γ-matrices contain the r = 2 Pauli matrices.

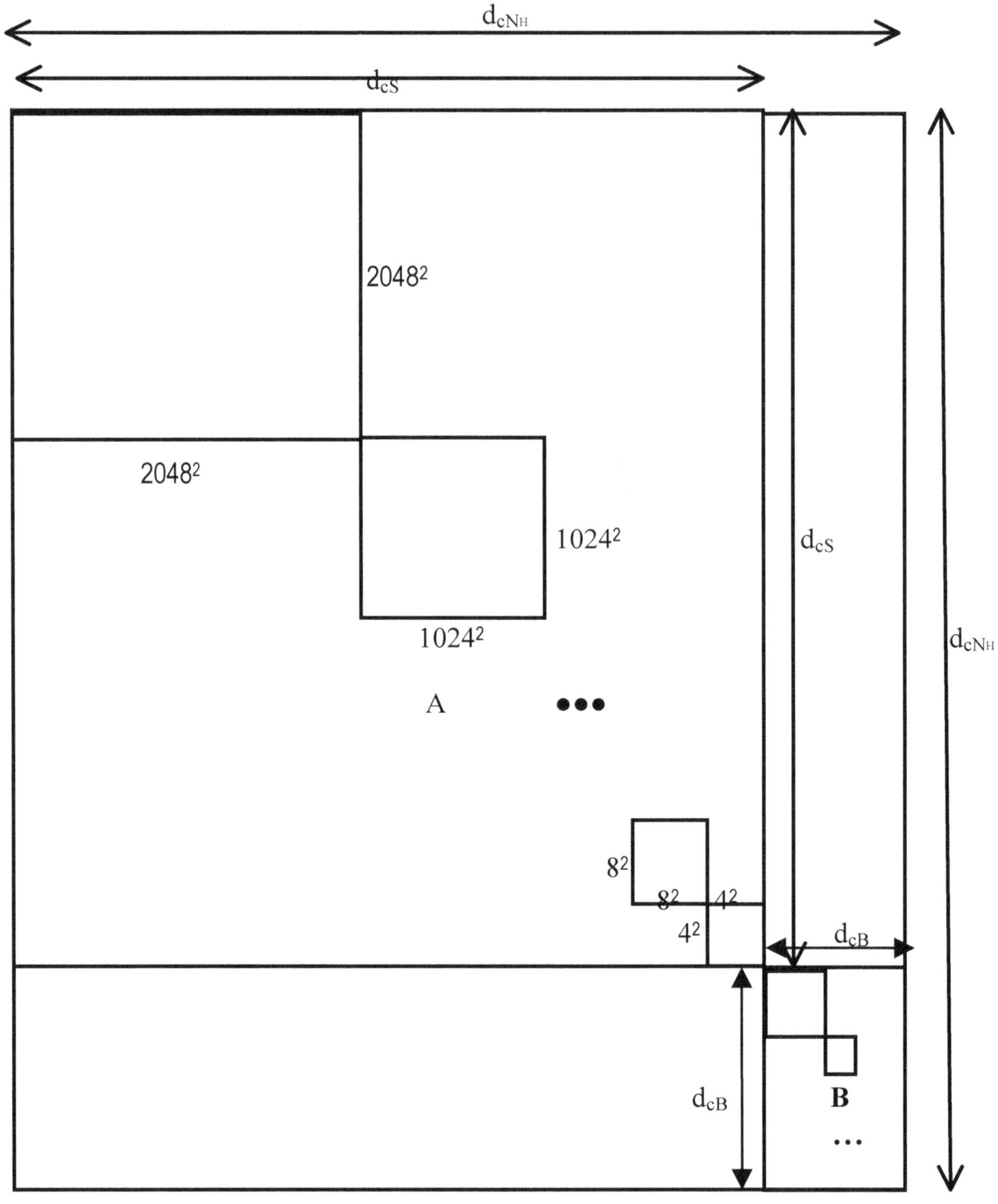

Figure 10.11. Form of a HyperUnification transformation. It is also the form of the dimension array of 42 dimension space-time. Ten HyperCosmos dimension array's column lengths are shown. Ten Second Kind HyperCosmos dimension array's column lengths are contained in the d_{cB} by d_{cB} block. The figure is not drawn to scale.See Chapter 6 of Blaha (2023a) for details.

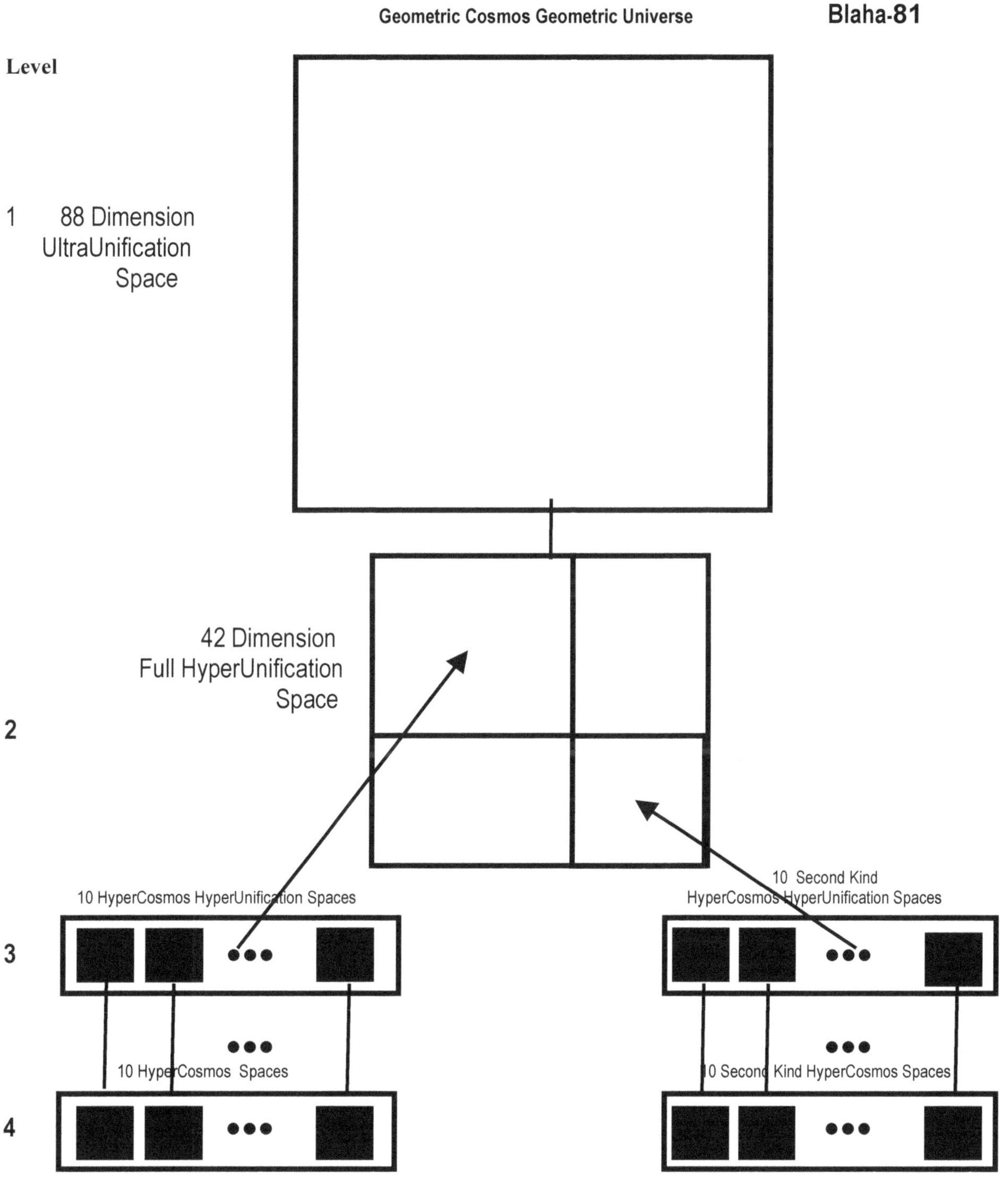

Figure 10.12. Diagram of the four levels of the Cosmos Spaces. They contain 42 spaces with no contributions from Limos spaces which are treated separately.

Consistency Condition for Cosmos Spaces

11. Origin of the Tenth Cosmos Space and Universe

We begin this book with a major extension of the consideration of the generation of the initial Parent universe. We will see that it leads to a set of expressions for important constants such as the origin of the power $2 \cong e\pi/4$; the choice of 10 Cosmos spaces; the choice of 18 space-time dimensions; the choice of $c_g = 0.0785 \cong \pi^2/128$ which appears in a universal relation $kT = mc_g$, as a new universal constant.

Remarkably, in addition, we find the Standard Model interaction coupling constants also have a powers sequence in 2 clearly suggesting a geometric origin of the Standard Model interactions (shown later). The values of the coupling constants, and other quantities are directly expressible in fundamental constants, π and e (the base of Natural logarithms), and simple integer powers of 2. The values of coupling constants depend on fermion nature (geometric) and geometry.

The conclusion of this work is:

**Geometry is the Basis of Cosmos Theory,
the Standard Model, and Internal symmetry Interactions.**

11.1 The Origin of the Cosmos

We initially chose an r = 18 space as the space of the original Parent universe raises several issues. (We now prove it.)

The logic of the origin point: There can be no time before the definition of the Parent space dimension array since no space or universe existed before its definition. Yet a Parent universe cannot exist without the definition of the Parent's dimension array. Thus the dimension array and Parent universe must be simultaneously defined and created.

The only way that one may reasonably resolve this state of affairs is to define both as the consequence of a consistency condition. We suggest the consistency condition for the minimum Parent universe dimension is the requirement that its virtual thermodynamic outward pressure equals the Casimir vacuum inward pressure from its virtual external vacuum Casimir pressure at the point of origin of the virtual universe.[39] See Fig. 11.1. Below this minimum dimension the Parent universe cannot exist since the virtual vacuum pressure would cause it to be contracted into non-existence.

The dimension r of the Parent universe fixes a maximum Physical Cosmos space dimension.[40] Within the Parent universe there are universes of lower dimension corresponding to lower dimension Cosmos spaces.

---

[39] Much of this chapter is based on Blaha (2022d) and (2023e).

[40] Depending on whether the Parent universe is mildly-expanding (r = 18) or expanding (r = 20). We discuss this issue later.

11.2 Balance Between Expansion and Contraction

When a possible virtual universe is "virtually" created it may expand (Hubble expansion) or it may contract to a point and then cease to exist, or it might be static. The decision is based on the dynamics of the virtual creation. Does the virtual universe possess sufficient mass-energy to generate a pressure for expansion or does the potential surrounding[41] virtual vacuum pressure causes the universe to contract and be "snuffed out"?

The balance between inward and outward pressure can be represented with

$$H = P - C \tag{11.1}$$

where P is the pressure of the energy-momentum contents of the virtual universe and C is the Casimir force due to the virtual vacuum energy embodied in the surface of the universe. There are three possibilities:

$H < 0$	Casimir force causes contraction to a point	(11.2)
$H = 0$	Balance of forces – "stability"	
$H > 0$	Pressure causes (Hubble) expansion of universe	

We wish to use H to determine the Cosmos Theory Physical spectrum. It has a lower limit at $r = 0$ but no apparent upper limit. We now view H as determining the maximum dimension r_{Parent} of the Physical Cosmos spectrum:

$H \geq 0$	"Normal" universe $r \leq r_{Parent}$	(11.3)
$H < 0$	Virtual universe lapses into non-existence	
$H = 0$	Static universe	

We now proceed to determine H for the space of dimension r.

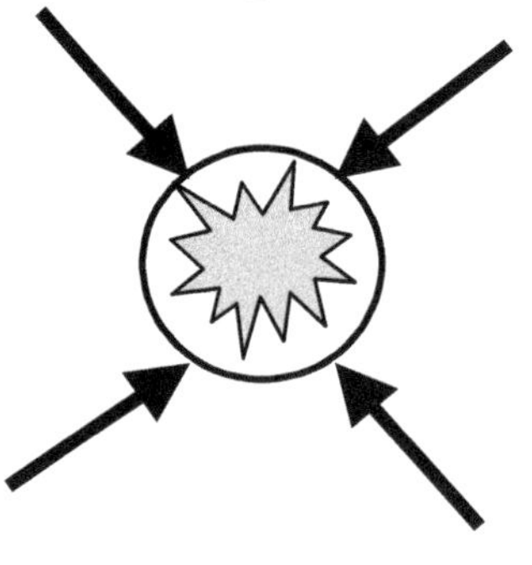

Figure 11.1. Schematic diagram of virtual universe's "vacuum" bubble showing the confining Casimir force of the exterior vacuum. This force is countered by internal fermion gas pressure symbolized by a jagged interior star.

[41] In the *surface* of the virtual universe.

11.3 Number of Fundamental Particle Species

Cosmos Theory specifies the number of fundamental fermion species N for a universe in a space of dimension r:

$$N = 2^{r+4} \tag{11.4}$$

The determination of the maximum number of fermions from H will fix the maximum dimension r.

11.4 Conditions at the Consistency Point

We now consider an infinitesimal box containing the energy-momentum of a virtual universe. The box contains fermions with the initial features:

1. All fermions are massless.
2. All interactions are zero.
3. All baryon numbers are zero.

11.5 Thermodynamic Outward Pressure

We base the thermodynamic pressure on the Ideal Gas Law[42]

$$P = NkT/V \tag{11.5}$$

where P is the pressure, T is the temperature, k is Boltzmann's constant, and N is the number of particles in unit volume V.

Assuming the fermion gas occupies a spherical spatial ball with spatial volume

$$V_{r-1} = \pi^{(r-1)/2}a^{(r-1)}/\Gamma((r-1)/2+1) \tag{11.6}$$

with radius a in an r dimension space-time:

$$P = N\,\Gamma((r-1)/2+1)kT/(\,\pi^{(r-1)/2}a^{(r-1)}) \tag{11.7}$$

where Γ(n) is the factorial function. The number of fermions N is given by eq. 11.4.

11.6 Casimir Effect Confining Force

This section calculates the force on a universe due to the Casimir effect.[43] We assume the universe is in a vacuum energy bubble in an r – 1 dimension spatial region of radius a. It generates a confining force on the surface of the universe. We start with the infinite Dirac fermion vacuum energy per unit volume:

---

[42] One might think of introducing more complex forms of gas laws based on the volume occupied by the gas particles and so on. However our assumption of zero interactions at the origin of universes eliminates the need for such "refinements." One might also thinking of introducing an attractive gravitation term in the consistency condition. Here again the assumption of no interactions at the origin point removes that concern.

[43] H. G. B. Casimir, Koninkl. Ned. Adak. Wetenschap. Proc. **51**, 793 (1948), S. K. Lamoreaux, Phys. Rev. Lett. **78**, 5 (1997), L. S. Brown and G. J. McClay, Phys. Rev. **184**, 1272 (1969), and references therein.

$$U_\infty = \int_0^\infty d^{r-1}k\,(k^2 + m^2)^{1/2} \qquad (11.8)$$

where m is the common fermion mass, which we take to be zero by section 11.4.

Due to the vacuum-less bubble, the vacuum energy changes to

$$U = U_\infty - U_{bubble} \qquad (11.9)$$

where the missing bubble vacuum energy per unit volume is

$$U_{bubble} = \int_d^\infty d^{r-1}k\,(k^2 + m^2)^{1/2} \qquad (11.10)$$

where d is the momentum space lower limit of the excluded vacuum energy within the bubble. The Casimir deficit is

$$U = \int_0^d d^{r-1}k\,k \qquad (11.11)$$

Assuming the momentum d corresponds to the spatial radius, a, of the bubble

$$a = 1/(mc) = 1/d \qquad (11.11a)$$

where c is the speed of light we find the energy deficit per unit volume is

$$U \cong \pi^{(r-1)/2} a^{-r}\,(r - 1)/[r\Gamma((r - 1)/2 + 1)] \qquad (11.12)$$

Multiplying by the volume of the bubble we find a total bubble energy:

$$U_{tot} \approx V_{r-1}\,U \qquad (11.13)$$

$$\approx \pi^{r-1} a^{-1}[(r - 1)/r]/\Gamma((r - 1)/2 + 1)^2 \qquad (11.14)$$

Then

$$\partial U_{tot}/\partial a \cong -\pi^{r-1} a^{-2}\,[(r - 1)/r]/\Gamma((r - 1)/2 + 1)^2 \qquad (11.15)$$

A bubble surface of radius a has the area

$$S_{r-2} = 2\pi^{(r-1)/2}\,a^{r-2}/\Gamma((r - 1)/2) \qquad (11.16)$$

The Casimir force on the surface is

$$C = (S_{r-2})^{-1}\,\partial U_{tot}/\partial a \qquad (11.17)$$

$$= -\tfrac{1}{2}\pi^{(r-1)/2} a^{-r} [(r - 1)/r] \Gamma((r - 1)/2)/\Gamma((r - 1)/2 + 1)^2$$
$$= - \pi^{(r-1)/2} a^{-r}/[r\Gamma((r - 1)/2 + 1)]$$

11.7 H Balance

We now set the outward pressure of the internal fermion "gas" equal to the inward Casimir vacuum force by eqs. 11.7 and 11.17:

$$H = P - C$$
$$= N\, \Gamma((r - 1)/2 + 1)kT/(\pi^{(r-1)/2}a^{(r-1)}) - \pi^{(r-1)/2} a^{-r}/[r\Gamma((r - 1)/2 + 1)] \quad (11.18)$$

Eliminating common factors in H we find the condition becomes

$$H' = N\, akT - \pi^{(r-1)}/[r\Gamma((r - 1)/2 + 1)^2] \quad (11.19)$$

At the critical point of eq. 11.2 we see H' = 0 implying

$$N = \pi^{(r-1)}/[arkT\, \Gamma((r - 1)/2 + 1)^2] \quad (11.20)$$

For collapse $H' < 0$

$$N < \pi^{(r-1)}/[arkT\, \Gamma((r - 1)/2 + 1)^2] \quad (11.21)$$

For expansion $H' > 0$

$$N > \pi^{(r-1)}/[arkT\, \Gamma((r - 1)/2 + 1)^2] \quad (11.22)$$

11.8 N and H' as Functions of Dimension r

We find that the choice of

$$akT = 2.635 \times 10^{-10} \quad (11.23)$$

and the Cosmos dimension array size choice $N = 2^{r+4}$, give a critical point at dimension r = 18.[44] Thus at minimum an r = 18 Parent universe can exist.

For even number dimensions below r = 18 (assuming eq. 11.23) we find H' is negative indicating such a Parent universe could not exist. The Casimir force prevents its existence. At r = 18 the pressure is equal to the Casimir force thus allowing an 18 dimension Parent universe within which child universes may be created. The r = 18 virtual universe is the minimal[45] (in r) universe, and thus the Parent universe.

On this basis we can choose a Physical set of ten Cosmos spaces. We can create one or more 18 dimension Parent universes. Each Parent universe is stationary. It balances expansion and contraction. A parent universe can contain universes of lesser Cosmos dimension since they have a less strong Casimir force with which to contend.

---

[44] Later in section 1.10 we refine the choice of critical dimension to r = 17.08 based on the critical turning point of the factor $(2e\pi/r)^r$ at r = 17.08 of $2e\pi/r$ from greater than 1 (contraction) to less than 1. A minimum r > 17.08 is required to avoid collapse to a point. We chose the next greater even integer r = 18 as the Parent space dimension.

[45] We require the dimension to be an even number.

If we wish the Parent universe to be more rapidly *expanding* then we can choose the r = 20 space's universe as the Parent universe. This choice would expand the Physical set of HyperCosmos spaces to eleven.

11.9 Support for the Power Series Form of Dimension Arrays

The size of dimension arrays N is a power series in 2 as shown in eq. 11.4. If we use eq. 11.20, and use Stirling's approximation for the gamma function, a power series emerges with exponent r for large values of r.

Stirling's approximation is

$$n! = (2\pi n)^{½}\ (n/e)^{n} \tag{11.24}$$

For large r >> 1, n ≅ r/2 and

$$\Gamma((r-1)/2 + 1) \rightarrow (\pi r)^{½}(r/2e)^{r/2}$$

Substituting in eq. 11.20 gives

$$\begin{aligned} N &= \pi^{(r-1)}/(arkT) \times 1/(\pi r(r/2e)^{r}) \\ &= (2e\pi/r)^{r}/(\pi arkT) \\ &= (17.08/r)^{r}/(\pi arkT) \end{aligned} \tag{11.25}$$

It shows the same form of power series with exponent r as N in eq. 11.4. The similarity in form, which is based on the form of volume and surface area in higher dimension spaces, reflects their joint basis in differential forms, n-forms and wedge expressions, seen before.

We see a strengthening of the view of dimension array sizes as based on geometry.

11.10 Why Ten Physical HyperCosmos Spaces?

An examination of eq. 11.25 shows that there is a more refined critical point value due to the factor

$$(2e\pi/r)^{r} \tag{11.26}$$

where the number 2eπ/r transitions from greater than one to less than one at r = 2eπ = 17.08. At r = 18 the Casimir force has changed from greater than the thermodynamic pressure to less than the thermodynamic pressure OR from contraction to expansion of the Parent universe. Thus r = 18 marks the beginning of an acceptable expanding Parent universe. At r < 18 universes contract to a point and thus could not generate a Parent universe. Parent universes can have expanding child universes within them.

The key scale for universe, and thus space dimension, is 2eπ = 8(eπ/4) ≅ 18. *We view this result as a confirmation of the approach in this chapter and the choice of r = 18 as the largest possible Physical HyperCosmos space as well as of the 10 Physical HyperCosmos space spectrum.*

11.11 Implications of Eq. 11.25

The consistency condition that we have found has a number of important results including the possibility of a new constant of nature c_g = 0.0785. This constant appears in many contexts in Elementary Particle theory and Astrophysics. See Blaha (2023e).

The implications of the consistency condition are:

1. The r = 18 space is the first even dimension integer (most minimal) space whose universe does not collapse. Thus it is an optimal choice for the parent space for an r = 18 universe. Within that universe sub-universes may be created that may undergo Hubble expansion. It is possible that there are many such unrelated parent universes with their own sub-universes.

2. Due to the factor $(2e\pi/r)^r$ the number 17.08 marks a critical dimension for the transition from a contracting universe.

3. The consistency condition embodies a factor $e\pi/4 = 2.13$ where e = 2.718 is the natural logarithm base constant. It appears in the factor

$$(2e\pi/r)^r = (8\times(e\pi/4)/r)^r = (8\times(e\pi/4)/r)^r \cong (17.08/r)^r \qquad (11.27)$$

4. Thus $e\pi/4 \cong 2$ is the constant in the dimension array d_{dr} series of powers to within 7% - a good result considering the approximate nature of the consistency condition calculation.

5. The value of the eq. 11.25 factor

$$akT$$

with the consistency condition value (eq. 11.23)

$$akT = 2.635 \times 10^{-10} \qquad (11.28)$$

has an important further interpretation when kT is approximated by[46]

$$kT = mc_g \qquad (11.29)$$

where

$$c_g = 0.0785 \qquad (11.30)$$

Using a = 1/mc (where c is the speed of light c = 3×10^8 m/s from eq. 11.11a) we find

$$akT = 0.0785/c = 2.62 \times 10^{-10} \qquad (11.31)$$

which differs by 2.4% from the consistency condition value above.

---

[46] The constant c_g appears in many contexts: elementary particles, universe studies, and Cosmos universes studies. See the next section for details.

6. When the $kT = mc_g$ relation is substituted in eq. 11.25 we find the consistency condition is independent of the mass m.

Coupling and Other Constants of Cosmos Theory

12. A New Constant of Nature c_g

The author's new constant c_g appears in many contexts in our studies in earlier books such as Blaha (2023e):

Blaha (2023e) *Cosmos Theory: The Sub-Particle Gambol Model*

Describes its use in deep inelastic lepton and quark resonances, production and interactions, hadron-hadron collisions; neutrino oscillations e - μ, μ – τ; particle stability pressure vs. Casimir force, and Hubble expansion. See section 9.8 and 9.9. In section 9.9 (eq. 13.22) we find an approximate relation, which we now express as

$$c_g = \pi^2/128 = 0.0771$$

giving a 2% difference from the approximate relation we found $c_g = 0.0785$.

Blaha(2024a) *Cosmos-Universe-Particle-Gambol Theory*

Here we found the following uses of c_g:

The Cosmos Theory Distribution of Universes
Anti-Planckian Model of Hubble Expansion Parameter
Our Universe as a Megaverse Particle
Universe-Megaverse Boundary
Casimir Force Limit on Hubble Expansion
S_8 Universe-Gambol Clumping Models
Variation of Universe Cluster Mass-Energy Sizes
Neutron Star Quark Core Gambol Model
GVDM Photon – ρ meson Gambol Model
Higgs and Massive Vector Boson Gambol Models

The close matches in values of akT = 2.635×10^{-10}, of kT = mc_g, and of c_g values before suggest it is a new universal constant.

13. A New Level of Physical Reality

The calculation of the consistency condition for Cosmos Theory has led to a number of simple expressions based on fundamental mathematical constants that raise the possibility of a new deeper level of Reality at the base of Cosmos Theory and Physics in general:

Quantity	Expression	Value	Known Value	Deviation
d_{dn} Power series Variable	$e\pi/4$	2.13	2	6.5%
c_g = kT/m constant	$\pi^2/128$	0.0771	0.0785	1.8%
Consistency Condition dimension	$8\ e\pi/4$	17.08		
akT	c_g/c	2.62×10^{-10}	2.635×10^{-10}	2.4%

To these values may be added a similar looking expression for α:

Fine Structure Constant[47] α	$e^2/1024$	0.00721438	0.0072973525643	1.15%

where e is the natural logarithm base e = 2.718. The approximation compares extremely well with the experimentally known value of α. The Fine structure constant appears as a constant in QED and ElectroWeak theory. *Its appearance above with an analogous form of expression suggests that it has a geometric origin.* The constant e that appears in the above calculations follows from Stirling's approximation for gamma functions appearing the expressions for higher dimension volumes and surface areas.

The relative closeness of the calculated value of "fine structure constant" to the experimentally known value is very encouraging.

The expression for the Fine Structure Constant above seems to be repeated for the other Standard Model interactions' coupling constants. We note factors of four between the known values suggesting a correspondence to the factors of four in Cosmos Theory. The calculated and experimental values below are based on powers of two is close to the known experimental values.[48]

---

[47] Calculated to known experimental value in the Johnson-Baker-Willey QED by this author. See Blaha (2019f) and references therein.

[48] The appearance of SU(4) is speculated. The interactions for SU(n) interactions for n > 4 are presently unknown. If such interactions exist then their strength might lead to total confinement such as we see for quarks.

	ESTIMATE		EXPERIMENT		
Interaction	**Expression**	$g^2/4\pi$ **Value**	**g**	**Known Value** $g^2/4\pi$	**Deviation**
U(0)	$e^2/4096$	0.0018036	-	-	-
U(1) α	$e^2/1024$	0.00721438	0.303	0.0072973525643	1.15%
SU(2) g	$e^2/256$	0.0289	0.63	0.0316	9.3%
SU(3) α_S	$e^2/64$	0.115	1.21	0.117	1.7%
SU(4)[49] ?	$e^2/16$	0.462	2.4?	0.458	0.087%

where g is the known coupling constant value[50] and e = 2.718 is the natural logarithm base. *Internal symmetry coupling constants have geometric values*[51].

There appears to be a sequence of powers in the expressions for the coupling constants:

$$\text{SU(n)} \qquad g(n)^2/4\pi = e^2/2^{12\,-2n} = 2^{\,r\,-\,12}\,e^2$$

$$g(n) = e\,(4\pi)^{½}\,2^{\,r/2\,-\,6} = 9.635\;2^{\,r/2\,-\,6} = 0.1505\;2^{\,r/2} \qquad (13.1)$$

where r = 2n is the number of *real* dimensions in the group's fundamental representation. The decrease in $g^2/4\pi$ by factors of four is analogous to the quadrupling effect in dimension array size in Cosmos Theory. *Thus interactions, and their strengths, appear to be geometric in origin.*

13.1 Weinberg Angle θ_W Calculation

The Weinberg angle is related to the electromagnetic and Weak coupling constants with

$$e^2/g^2 = \sin^2\theta_W$$

where e is the electric charge. From the above table, using our calculated values, we find

$$\sin^2\theta_W = 256/1024 = 0.25$$
$$\sin\theta_W = 0.5$$

compared to the experimental value of $\sin^2\theta_W = 0.231$. An accuracy to 8% that further supports the power sequence of eq. 13.1.

Since the Weinberg angle is related to massive vector boson masses we see the conventional definition gives

$$m_W/m_Z = \cos\theta_W = 0.866$$

---

[49] This value is based on the "doubling trend" seen in the three known coupling constants above.

[50] All coupling constant values are based on data from Particle Data Group 2022.

[51] We add a U(0) line for the sake of completeness.

in our approximation.

The Coupling Constants, the dimension arrays, and the relation of coupling constants to masses (and thereby to energies) all reflect their common interpretation as manifestations of degrees of freedom. There is a unifying principle at the basis of Cosmos Theory in its spaces, symmetries, universes, universe structure, energies, and particle interactions.

14. Coupling Constants as Dimensions

Coupling constants (excepting gravitation) are numbers. We have shown that they have the same sequence of powers as we saw in dimension arrays.

We can treat the dimension array of dimension r with the size:

$$d_{dr} = 2^{r+4}$$

as corresponding to a special unitary group, which is the largest possible group in dimension r,

$$\text{MAX} = \text{SU}(2^{r+3})$$

based on the number of dimensions in the group's fundamental representation.

From the previous expression for $g^2/4\pi$, the dimension array size in ratio to the coupling constant is

$$(g_{max}(n)^2/4\pi)/d_{dr} = 2^{r-12}\, e^2/2^{r+4} = 2^{-16}e^2$$

in dimension r where $n = d_{dr}/2$. Thus d_{dr} dimensions are required for $\text{SU}(2^{r+3})$ and

$$d_{dr} = 2^{16} g_{max}{}^2/(4\pi e^2) \qquad (14.1)$$

One dimension has the coupling constant value

$$2^{12-r} g_{max}{}^2/(4\pi e^2) \qquad (14.2)$$

We now have a direct relation in dimension r between the dimensions in a dimension array and the size and coupling constant of the maximal group in dimension r.

14.1 Dimension Cost of an Interaction Coupling Constant

Each interaction has an associated coupling constant, which is related to the group of the interaction:

$$\text{SU}(n) \qquad g(n)^2/4\pi = e^2/2^{12-2n}$$

If we ask what the "cost" is, in dimensions, of a coupling constant we find

$$\text{Cost in dimensions} = 2^{2n-12}\, e^2/2n \ \text{ per dimension}$$

14.2 Coupling Constants as Products of Dimension Degrees of Freedom

We have shown coupling constants appear to form a powers sequence. We now consider the expression of a coupling constant as the product of a number of factors that are based on geometry.

First we note that n is the number of fermions in a fundamental representation of SU(n). The coupling constant g has the form:

$$g(n) = e\,(4\pi)^{1/2}\,2^{n-6} = e\,(2^{-10}\pi)^{1/2}\,2^{n} = 0.1505\;2^{n} \qquad (14.3)$$
$$= g_g\,(\text{number of spin states per fermion})^{\text{number of fermions}}$$

The coupling constant is a fundamental value times the number of spin states (a geometric property of fermions equal to half the column length of a spinor) raised to the number of fermions.

We see that g(n) has n factors – one factor for 2 spin states for each fermion in the SU(n) fundamental representation. The origin of

$$g_g = e\,(2^{-10}\pi)^{1/2} \qquad (14.4)$$

is not fully understood due to the appearance of the natural logarithm base e.

Eq. 14.3 implies that the coupling g(n) for SU(n) is g_g times n factors of 2 which represents the number of spin states per fermion in the SU(n) fundamental representation in four dimensions. Thus it is a product of degrees of freedom.

An important point is the number of factors is equal to the number of fermions in the SU(n) fundamental representation. *The coupling constant "felt" by one fermion depends on the total number of fermions in the group's fundamental representation.*

14.3 A Possible Generalization to Higher Space-time Dimensions

Note g_g has the form

$$g_g = e\,(2^{-12}\,S_2)^{1/2} \qquad (14.5)$$

where S_2 is the surface area of a unit 3-sphere in four dimensions.

We can generalize eq. 14.3 to higher dimensions using the column length of higher space-time dimensions r:

$$g^r(n) = e\,(2^{-12}S_{r-2})^{1/2}\,(2^{r/4})^{n} = e\,(2^{-12}S_{r-2})^{1/2}\,2^{nr/4} \qquad (14.6)$$
$$= g^r{}_g(n)\,2^{nr/4}$$

using eq. 11.16 for the surface of a unit (r – 1)–sphere in r dimensions.

The constants $g^r{}_g(n)$ have the form

$$g^r{}_g(n) = e\,(2^{-12}\,S_{r-2})^{1/2} \qquad (14.7)$$

They are geometric. The factor of e is discussed in the following section.

14.4 Form of $g^r{}_g(n)$

We can expand

$$g^r{}_g(n) = e\,(2^{-12}\,S_{r-2})^{1/2}$$

using the surface area S_{r-2} of eq. 11.16 and Stirling's approximation for $\Gamma(n)$ in eq. 11.24.

We find the approximate expression in terms of e and π:

$$g^r{}_g(n) \sim e^{5/4}\,(e\pi)^{(r-2)/4}\,R$$

where R is a numeric expression of integers. The factor of e appears to have an origin other than geometry. It is possibly a result of Quantum Perturbation Theory. The appearance of the eπ factor is interesting in light of its appearance in expressions for constants seen earlier.

15. Inclusion of Gravitational Coupling Constant G

The gravitational coupling constant G of our universe can be put in the coupling constant framework that we established. We augment the Coupling Constant table to the table of Fig. 5.1.

	ESTIMATE		EXPERIMENT		
Interaction	**Expression**	$g^2/4\pi$ **Value**	**g**	**Known[52] Value** $g^2/4\pi$	**Deviation**
U(0) α_0	$e^2/4096$	0.0018036	-	-	-
U(1) $\alpha_1 = \alpha$	$e^2/1024$	0.00721438	0.303	0.0072973525643	1.15%
SU(2) $\alpha_2 = g^2/4\pi$	$e^2/256$	0.0289	0.63	0.0316	9.3%
SU(3) $\alpha_3 = \alpha_S$	$e^2/64$	0.115	1.21	0.117	1.7%
SU(4)[53] α_4	$e^2/16$	0.462	2.4?	0.458	0.087%
•					
$\alpha_n = g_n^2/4\pi = e^2\, 2^{2n-12}$					
•					
•					
SU(256) α_{256}	$e^2 2^{500}$	2.418×10^{151}	1.74×10^{76}	-	-

Figure 15.1. Extended Coupling Constants table.

SU(256) has an associated coupling constant:

$$\alpha_{256} = g^2/4\pi = e^2 2^{500} \tag{15.1}$$

We now find the mass associated with gravitational coupling:

$$G = g_{256}{}^2/4\pi\ M^{-2} = 0.67 \times 10^{-38}\ (\text{GeV}/c^2)^{-2} \tag{15.2}$$

We set

$$g_{256}{}^2/4\pi = \alpha_{256}$$

since the 256 fundamental fermions (both normal and Dark) in UST correspond to a SU(256) symmetry group. Then

$$M^2 = (g_{256}{}^2/4\pi)/G = 2.418\times10^{151}/(0.67\times10^{-38}) = 3.61 \times 10^{189}\ (\text{GeV}/c^2)^2$$

and

[52] All coupling constant values are based on data from Particle Data Group 2022.
[53] This value is based on the "doubling trend" seen in the three known coupling constants g above.

$$M = 6 \times 10^{94}\ \text{GeV/c}^2$$

This value of M is a factor of 3.5×10^{13} times the estimated total[54] mass of the *visible* universe:[55]

$$M_{\text{observable}} = 1.71 \times 10^{81}\ \text{GeV/c}^2 = 2^{269.85}\ \text{GeV/c}^2$$

This result suggests that the universe is a form of particle with mass M and symmetry group SU(256). It also suggests the universe is much larger than the observable part. We have suggested this possibility several times previously in other contexts.

SU(256) like SU(3) might have been associated with confinement. However this possibility is removed by the mass factor M^{-2} that is required on dimensional grounds in gravitational dynamic equations.[56] This factor causes the gravitation G to be weak.

We appear to have a viable estimate of the value of G and its origin in a coupling constant formulation.

Later we extend this discussion to the case of all universes.

Note: if the symmetry group was SU(211) then one would obtain the current value of the total observable universe mass.

---

[54] This value is for normal mass.

[55] There are many suggestions of a universe mass-energy much greater than the observable.

[56] Models which change gravity from 1/r Newtonian gravity usually change the space-time dependence – not G.

Basis of Current Masses in Cosmos Theory

16. Current Masses in Cosmos Theory and The Standard Model

In Blaha (2024f) we developed a theory of coupling constants that yielded the known coupling constants of The Standard Model together with extremely accurate mathematical expressions for each of the known interactions. These expressions were shown to have a basic form within a Cosmos Theory perspective as a constant e (the base of natural logarithms) times the number of degrees of freedom (spins) for the fermions of the fundamental representation of each coupling constant's internal symmetry group. Thus they have a fundamental basis.

We now turn to the other known set of constants in The Standard Model and Cosmos Theory – the eight fermion (quark and lepton) masses of the first generation of fermions. We will show that fairly simple patterns exist in these fermion masses that lend themselves to being expressed as simple mathematical formulas. The form of these formulas partly reflects the form of previously derived coupling constants' values.

The fermions of the fermion mass spectrum, so derived, express the masses in the form of integer mathematical factors times powers of π. In a sense the appearance of π in masses complements the appearance of e in coupling constant values. Note that $e\pi/4$ is approximately 2, which is the power sequence variable in the dimension array sizes (2^{r+4}) of Cosmos Theory spaces. Thus masses and Cosmos Theory dimensions have similar expressions.

We begin with some of the coupling constants of Fig. 15.1 (found in Blaha (2024f)).

		ESTIMATE			EXPERIMENT	
Interaction		**Expression**	$g^2/4\pi$ **Value**	**g**	**Known**[57] **Value** $g^2/4\pi$	**Deviation**
U(0)	α_0	$e^2/4096$	0.0018036	-	-	-
U(1)	$\alpha_1 = \alpha$	$e^2/1024$	0.00721438	0.303	0.0072973525643	1.15%
SU(2)	$\alpha_2 = g^2/4\pi$	$e^2/256$	0.0289	0.63	0.0316	9.3%
SU(3)	$\alpha_3 = \alpha_S$	$e^2/64$	0.115	1.21	0.117	1.7%
SU(4)[58]	α_4?	$e^2/16$	0.462	2.4?	0.458	0.087%

Figure 16.1. The Coupling Constants for first generation internal symmetries in Cosmos Theory and The Standard Model.

Note the interrelationships embodied in the coupling constant expressions such as:

[57] All coupling constant values are based on data from Particle Data Group 2022.

[58] This value is based on the "doubling trend" seen in the three known coupling constants above.

$$\alpha_4{}^2 = e^2\alpha_2 \tag{16.1}$$

16.1 Origin of U(0)

The U(0) group has the two elements ±I. It might appear in Lagrangian mass terms such as

$$g_0\,\bar{\psi}_i U(0) m_{ij} \psi_j \tag{16.2}$$

where $i < j$ and $i, j = \nu_e, \nu_\mu, \nu_\tau, \nu_{\tau'}$.

16.2 Fermion Generations

The Standard Model *defines* fermion generations (Fig. 16.2). Cosmos Theory and, in particular, The Unified SuperStandard Theory (UST) within it define generations in a manner based on the form of the r = 4 dimension array of the Cosmos Spectrum.[59] (Fig. 16.3) The UST has four layers, which each contain four generations.[60] Each generation has eight "normal" fundamental fermions. (There is also a corresponding eight Dark fundamental fermions in the Dark matter sector.) Thus there is a total of 4×4×8 = 128 normal fermions and 128 Dark fermions.

Standard Model Generations

Generation:	1	2	3
	u	c	t
	d	s	b
	e	μ	τ
	ν_e	ν_μ	ν_τ

Figure 16.2. The Standard Model generations.

We will consider the *Current* mass spectrum of the eight first generation "normal" fermions of UST. See Figs. 16.3 and 16.4. We will show they exhibit interesting regularities, which support the UST (and Cosmos) view of generations and also provide a simple, surprising, numerical fermion spectrum of current masses.

[59] See Blaha (2023d) and (2018e) for details.

[60] The four layers support the author's Layer groups and the four generatioons per layer support the author's Generation groups. There are also Connection groups. See Appendices 25-A and 25-B for a description of these groups, all of which were originally introduced some time ago by the author.

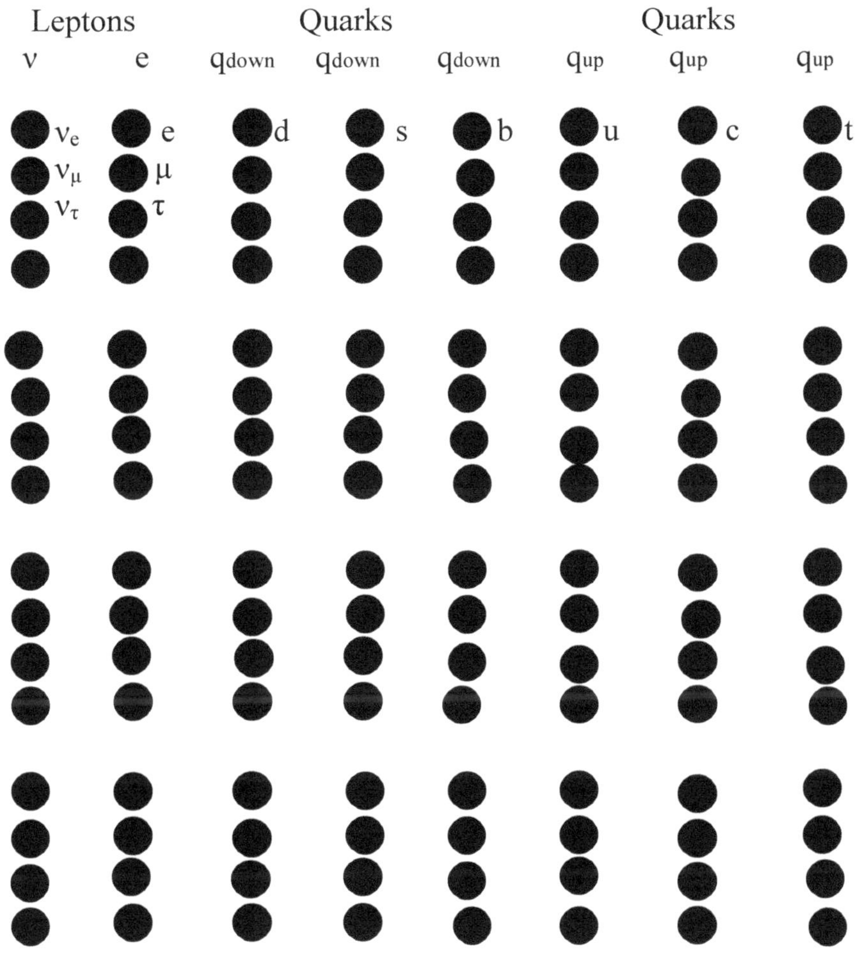

Figure 16.3. Four UST (and Cosmos) layers with each consisting of four generations of eight normal and eight Dark fermions. From Blaha (2023d) and the author's earlier books.

$\nu_e = 3.878 \times 10^{-11}$ GeV/c^2

Our ν_e mass estimate = .03878 ev/c^2 is consistent with known data estimates.

$\nu' = \nu_e\ (\alpha_0{}^2)^{-1} = 0.03878$ ev/c^2 $\times\ 4096^2/e^2 = 8.8 \times 10^{-5}$ GeV/c^2 where e = 2.718.

We scale ν_e by $(\alpha_0{}^2)^{-1}$. This scaling puts the neutrino mass on the same footing as the other masses. It may reflect the need to scale the neutrino mass by the U(0) "symmetry." Later we suggest a further need to scale mass values based on the ratio of SU(2) mass effects to SU(3) mass effects, which appear to vary as a function of fermion mass.

e $= 0.511 \times 10^{-3}$ GeV/c^2 = electron mass

u 1.76 mev/c2 = 1.76×10^{-3} GeV/c^2

We scale the u mass by a factor of 4/5 – again to put it on a similar footing as the other masses. Note: the u mass is 2.2 +0.5 -0.4 $\times\ 10^{-3}$ GeV/c^2 The 1.76 value is consistent with the lower bound 1.8.

c 1.27 GeV/c^2

t 172.76 GeV/c^2

d 3.76×10^{-3} GeV/c^2

We scale the d mass by a factor of 4/5 – again to put it on a similar footing as the other masses. Note: the d mass is 4.7 +.5 - .3 $\times\ 10^{-3}$ GeV/c^2 The 3.76 value is consistent with the lower bound 4.4.

s 95×10^{-3} GeV/c^2

b 4.18 GeV/c^2

Figure 16.4. Fundamental first generation current fermion masses. Each fundamental fermion symbol represents its mass.

17. Analysis of the First Generation Fermion Current Masses

At first glance the first generation mass spectrum of Fig. 16.4 does not appear to exhibit any regularities. In this chapter we will develop a new view of the spectrum based on the diagram in Fig. 17.1.

This diagram portrays the fermion masses on two levels based on the ElectroWeak separation of fermions in pairs. On each level the masses are ordered in increasing value. The diagram inverts the u and d level locations based on the overall ordering of top level masses as above the corresponding low level masses. Each ElectroWeak pair has the more massive above the less massive.

We introduce vertical arrows pointing to the lower mass of corresponding pairs. We introduce "angled" arrows between consecutive pairs pointing to the larger mass of adjacent pairs. This procedure generates a path between masses that leads to surprising regularities in fermion mass relations.

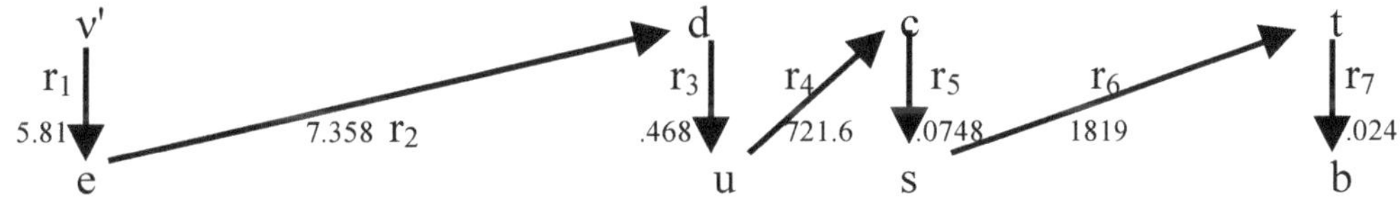

Figure 17.1. Fundamental first generation current fermion masses. Each fundamental fermion symbol represents its mass.

Fig. 17.1 displays ratios denoted r_i where i =1, 2, … , 7 that we will now define and use to relate ratios for the purpose of generating relationships between masses. We define the ratios with their corresponding numeric values obtained directly from Fig. 16.4.

$$r_1 = e/\nu' = .131/.02254 = 5.81 \qquad (17.1)$$
$$r_2 = d/e = 7.358$$
$$r_3 = u/d = 0.468$$
$$r_4 = c/u = 721.6$$
$$r_5 = s/c = 74.8\times10^{-3} = 0.0748$$
$$r_6 = t/s = 1.819\times10^3 = 1819$$
$$r_7 = b/t = 0.024$$

We can use these ratios and values to define fermion masses. Note first

$$\nu' = \nu_e\ (\alpha_0{}^2)^{-1} \qquad (17.2)$$
$$r_2/r_1 = 1.27 \cong 4/\pi$$

Then we use eq. 17.1 to obtain

$$
\begin{aligned}
e &= r_1\ \nu' \\
d &= r_2 e = r_2 r_1 \nu' \\
u &= r_3 d = r_3 r_2 r_1 \nu' \\
c &= r_4 u = r_4 r_3 r_2 r_1 \nu' \\
s &= r_5 c = r_5 r_4 r_3 r_2 r_1 \nu' \\
t &= r_6 s = r_6 r_5 r_4 r_3 r_2 r_1 \nu' \\
b &= r_7 t = r_7 r_6 r_5 r_4 r_3 r_2 r_1 \nu'
\end{aligned}
\qquad (17.3)
$$

This buildup is in "slice" factors of *mass* in a manner analogous to the buildup of coupling constants with spin degrees of freedom factors in section 4.2 of Blaha (2024f).

Following the arrows in Fig. 17.1 we see two patterns emerge. The set of angled arrows exhibit an increasing sequence of ratios. We form ratios of ratios and see approximate factors containing π appear:

$$
\begin{array}{ccccc}
r_2 & & r_4 & & r_6 \\
7.358 & & 721.6 & & 1819 \\
 & r_4/r_2 = 98.1 & & r_6/r_4 = 2.52 & \\
 & \cong 32\pi & & \cong \pi &
\end{array}
\qquad (17.4)
$$

The set of vertical arrows exhibit a decreasing sequence of ratios:

$$
\begin{array}{ccccccc}
r_1 & & r_3 & & r_5 & & r_7 \\
5.81 & & 0.468 & & 0.0748 & & 0.024 \\
 & r_1/r_3 = 12.42 & & r_3/r_5 = 6.28 & & r_5/r_7 = 3.116 & \\
 & \cong 4\pi & & \cong 2\pi & & \cong \pi &
\end{array}
\qquad (17.5)
$$

The angled line data gives:

$$r_4 = 32\pi r_2 \qquad r_6 = \pi r_4 = 32\pi^2 r_2 \qquad (17.6)$$

Note the powers of $\pi^{(i-2)/2}$ for i = 4, 6.
The vertical line data gives:

$$r_1 = 5.81$$

$$r_5 = \pi\ r_7 \qquad r_3 = 2\pi r_5 \qquad r_1 = 4\pi r_3 \qquad (17.7)$$

Using the above relations we find

$$r_1 = e/ v' = .131/.02254 = 5.81 \cong 2\pi \tag{17.8}$$
$$r_2 \cong 4/\pi\ r_1 \cong 8$$
$$r_3 \cong r_1/(4\pi) \cong 1/2$$
$$r_4 \cong 256\pi$$
$$r_5 \cong 1/(4\pi)$$
$$r_6 \cong 256\pi^2$$
$$r_7 \cong 1/(4\pi^2)$$

with $v' = 8.8 \times 10^{-5}$ GeV/c^2. Fig. 17.2 shows the resulting approximate mass values.

INITIAL ESTIMATE	DEVIATION RATIO	DATA
$e = r_1 v' = 2\pi v' = .553 \times 10^{-3}$	1.08	0.511×10^{-3} GeV/c^2
$d = r_2 e = r_2 r_1 v' = 2^4 \pi v' = 4.42 \times 10^{-3}$	1.18	3.76×10^{-3} GeV/c^2
$u = r_3 d = r_3 r_2 r_1 v' = 2^3 \pi v' = 2.51 \times 10^{-3}$	1.42	1.76×10^{-3} GeV/c^2
$c = r_4 u = r_4 r_3 r_2 r_1 v' = 2^{11} \pi^2 v' = 1.78$	1.4	1.27 GeV/c^2
$s = r_5 c = r_5 r_4 r_3 r_2 r_1 v' = 2^9 \pi v' = 141.55 \times 10^{-3}$	1.49	95×10^{-3} GeV/c^2
$t = r_6 s = r_6 r_5 r_4 r_3 r_2 r_1 v' = 2^{17} \pi^3 v' = 357.64$	2.07	172.76 GeV/c^2
$b = r_7 t = r_7 r_6 r_5 r_4 r_3 r_2 r_1 v' = 2^{15} \pi v' = 9.06$	2.16	4.18 GeV/c^2

Figure 17.2. Approximate estimates of the fermion current masses. The estimates can be viewed as approximations to the experimental data based on powers of 2 and π.

The agreement of the estimates[61] formed from directly examining the data, and the actual data, is good for e and d; is less good for u, c, and s; and much less good for t and b. We remedy these anomalies in chapter 18.

[61] The limitation to powers of 2 and π times v' makes the estimates non-trivial.

18. Revised Estimate of Current Fermion Masses based on SU(2)/SU(3) Variation

In the preceding chapter we developed the forms of eight known fundamental fermion masses based on a constant times a power of 2 and a power of π. These forms are analogous to the forms of coupling constants that we recently found based on powers of 2 and e – the base of the natural logarithms. Together this harmony of mass and coupling constant forms lends indirect support to Cosmos Theory. It is somewhat remarkable that coupling constants are based on e and masses are based on π. They are not jointly based on both. It suggests there is a deeper, new layer of particle interactions which we attempt to elucidate later.

We now turn to the possibility of improving the mass estimates of chapter 17 by the introduction of a factor giving a sliding scale of mass dependence in the ratio of an ElectroWeak SU(2) factor and a Strong Interaction SU(3) factor. We assume they contribute with equal strength to the e and d masses; with a mild dominance of ElectroWeak over Strong for u, c, and s quarks; and with dominance of Strong over ElectroWeak by a factor of 2 for t and b quarks.

We denote the ratio of the interaction strengths with

$$\alpha/\beta \equiv \alpha_{SU(2)}(m)/\beta_{SU(3)}(m) \qquad (18.1)$$

where m is the mass of the relevant fermion. We note that the ratio of 2 for t and b reflects the ratio of the coupling constants (1.21/0.63 = 1.92 ≅ 2) in Fig. 1.1. The value of α/β nicely progresses from 1 to ¾ to ½ as the fermion mass increases.

α/β	**REVISED ESTIMATE**	**DEVIATION RATIO**	**DATA**
1	$\nu_e = 0.0372$ ev/c^2 $= 3.72 \times 10^{-11}$ GeV/c^2		
1	$\nu' = \nu_e (\alpha_0^2)^{-1} = 8.4 \times 10^{-5}$ GeV/c$^2 \cong e2^{-15}$ GeV/c^2		
1	$e = r_1 \nu' = 2\pi \nu' = 0.511 \times 10^{-3}$	1.	0.511×10^{-3} GeV/c^2
1	$d = r_2 e = r_2 r_1 \nu' = 16\pi \nu' = 4.24 \times 10^{-3}$	1.13	3.76×10^{-3} GeV/c^2
¾	$u = r_3 d = r_3 r_2 r_1 \nu' = ¾ \times 2^3 \pi \nu' = 1.80 \times 10^{-3}$	1.02	1.76×10^{-3} GeV/c^2
¾	$c = r_4 u = r_4 r_3 r_2 r_1 \nu' = ¾ \times 2^{11} \pi^2 \nu' = 1.28$	1.001	1.27 GeV/c^2
¾	$s = r_5 c = r_5 r_4 r_3 r_2 r_1 \nu' = ¾ \times 2^9 \pi \nu' = 102 \times 10^{-3}$	1.07	95×10^{-3} GeV/c^2
½	$t = r_6 s = r_6 r_5 r_4 r_3 r_2 r_1 \nu' = 2^{16} \pi^3 \nu' = 171$	0.99	172.76 GeV/c^2
½	$b = r_7 t = r_7 r_6 r_5 r_4 r_3 r_2 r_1 \nu' = 2^{14} \pi \nu' = 4.34$	1.04	4.18 GeV/c^2

Figure 18.1. Revised fermion masses based on α/β and on the known value of e of 0.511 from which the values of v' and v are extracted and used to calculate the other fermion masses. The d discrepancy is the largest reflecting, perhaps, some unceratinty in the experiment value of d. Otherwise there is excellent agreement with experiment!

Strong SU(3) dominates at high mass; ElectroWeak SU(2) dominates at low mass. All estimates in Fig. 18.1 are close to the experimental values. Thus we have a representation of fermion masses in terms of powers of 2, powers of π, and the value of v' - comparable to coupling constants where coupling constants are powers of 2 multiplied by e – the base of natural logarithms. This suggests that Cosmos Theory is at the base of both coupling constant and fermion mass values. Why one uses π and e, and the other uses only e - the base of natural logarithms, will be shown to be the result of Geometry later.

Substituting $v' \cong e2^{-15}$ GeV/c^2 gives Fig. 18.2.

α/β	FERMION
1	$\nu_e = 0.0372$ ev/c^2 $= 3.72 \times 10^{-11}$ GeV/c^2
1	$\nu_e = \alpha_0^2\, v' = e^5 2^{-39}$ GeV/c^2
1	$v' = \nu_e\, (\alpha_0^2)^{-1} \cong e2^{-15}$ GeV/c^2
1	$e = r_1\, v' = \pi e 2^{-14}$ GeV/c^2
1	$d = r_2 e = r_2 r_1 v' = \pi e 2^{-11}$ GeV/c^2
¾	$u = r_3 d = r_3 r_2 r_1 v' = ¾ \times \pi e 2^{-12}$ GeV/c^2
¾	$c = r_4 u = r_4 r_3 r_2 r_1 v' = ¾ \times \pi^2 e 2^{-4}$ GeV/c^2
¾	$s = r_5 c = r_5 r_4 r_3 r_2 r_1 v' = ¾ \times \pi e 2^{-6}$ GeV/c^2
½	$t = r_6 s = r_6 r_5 r_4 r_3 r_2 r_1 v' = 2\pi^3 e$ GeV/c^2
½	$b = r_7 t = r_7 r_6 r_5 r_4 r_3 r_2 r_1 v' = \pi e 2^{-1}$ GeV/c^2

Figure 18.2. Revised form of fermion masses using $v' \cong e2^{-15}$ GeV/c^2.

18.1 Fermion Masses are Coupling Constants

A comparison of Figs. 18.2 and 18.3 shows a significant resemblance which we take to imply that fermion mass is analogous to a coupling constant. Note fermion masses couple the large and small parts of spinors. Note the appearance of e, the base of natural logarithms. Note also the powers of two. Masses differ from coupling constants in having powers of π.

Interaction		$g = (4\pi\, \alpha)^{½}$
U(0)		$g_0 = e2^{-6}$
U(1)	$\alpha_1 = \alpha$	$g_1 = e2^{-5}$
SU(2)	$\alpha_2 = g^2/4\pi$	$g_2 = e2^{-4}$
SU(3)	$\alpha_3 = \alpha_S$	$g_3 = e2^{-3}$
SU(4)	α_4	$g_4 = e2^{-2}$
		.
		.
		.

Figure 18.3. Coupling constants.

18.2 Two Sequences of Fermion Masses

We have found two sequences of fermion masses apparent in Fig. 18.1:

	SEQUENCE 1							SEQUENCE 2						
	e		u		c		t	ν'		d		s		b
Mass:	0.511×10^{-3}		1.80×10^{-3}		1.28		171	8.4×10^{-5}		4.24×10^{-3}		102×10^{-3}		4.34
Multiplier:		$2^5\pi$		$2^5\pi$		$2^5\pi$			32		32		32	

where the multiplier connects adjacent masses in each sequence separately.

We suggest all four generations of fermions implement this separation into sequences. See chapter 20.

The two sequences implement the multipliers:

Up type fermion sequence multiplier: $2^5\pi = 32\pi$
Down type fermion sequence multiplier: $2^5 = 32$

We believe the multiplier of $2^5 = 32$ appearing in both sequences is the number of fermions in a UST layer based on the concept that all fermions of a layer are intertwined. There are 32 normal fermions in a UST layer.

The explanation of the factors of π reflects internal universe shells that we will discuss later in chapter 23 – Geometry.

Note each mass has implicit e and π factors due to

$$\nu'_{new} = \nu_{e\ new}\,(\alpha_0{}^2)^{-1} \cong e\pi\ \times 10^{-5}\ \text{GeV/c}^2 \tag{18.2}$$

18.3 Factoring Mass Values in the Manner of Gambol Theory

Coupling constant values depend on powers of 2 with the power of 2 determined by the count of spin degrees of freedom of the fermions in the group's fundamental representation. Based on this consideration it is reasonable to consider factoring mass values in a manner similar to the factoring of particles into gambols. Gambol Theory defines the gambol mass m_g of a particle with a similar factoring equation:

$$m_g = M/s \tag{18.3}$$

where s is a power of 2 and M is a fermion mass, as we saw in our earlier books.

We now can consider factoring a mass such as the u quark mass into gambol masses. For example we can factor the u quark into eight parts with

$$m_g = ¾ \times \pi e 2^{-12}/2^8 == ¾ \times \pi e 2^{-20} \tag{18.4}$$

and proceed to consider applications in the Gambol Theory of quarks.

Factoring the masses of fermions as seen in Figs. 18.2 and 18.3 thus leads to a deeper view of fundamental fermions as having a type of conceptual partitioning.

19. Vector Boson Masses

The calculation of fermion mass data as functions of powers of 2 and π may be extended to the case of the vector boson masses in ElectroWeak theory. We previously showed the Weinberg angle θ_W, which is defined by

$$e^2/g^2 = \sin^2 \theta_W \tag{19.1}$$

using coupling constant values sets

$$\sin \theta_W = 0.5 \tag{19.2}$$

in close approximation to the known value of θ_W (where $\sin^2 \theta_W = 0.231$).

The W and Z masses are related by

$$m_W{}^2/m_Z{}^2 = \cos^2 \theta_W = 0.75 \tag{19.3}$$

They are known to have the values

$$\begin{aligned} W &= 80.377 \\ Z &= 91.19 \end{aligned} \tag{19.4}$$

where we use their symbols to represent their masses.

Noting that the t quark mass numerically satisfies

$$t/Z = 178.8/91.19 = 1.96 \cong 2 \tag{19.5}$$

we can use the Fig. 18.1 value of t to write

$$Z = t/2 = 2^{15}\pi^3 \, v' \tag{19.6}$$

thus giving an approximate expression for Z in terms of powers of 2 and π. Eq. 19.4 implies the W mass

$$W^2 = 0.75Z^2 = {}^3\!/\!_4 \; 2^{30}\pi^6 \, v'^2 \tag{19.7}$$

or

$$W = \sqrt{{}^3\!/\!_4} \; 2^{15}\pi^3 \, v' \tag{19.8}$$

Thus the W and Z masses have simple representations in terms of powers of 2 and π analogous to those of the fermion masses. There is a similar relation for the known Higgs particle mass.

20. Generations of Fermions

We have suggested that the Cosmos Theory portrayal of generations of fermions (Fig. 16.3) should be preferred especially in view of the patterns of first generation current fermion masses. (Fig. 18.1)

We now consider the extension of the first generation mass patterns and spectrum to the second and third generations. The masses of the μ and τ leptons (of the second and third generations) are known. We can use their mass ratios to *approximately* extend the quark and neutrino masses to the second and third generations.

The μ and τ mass ratios can be approximated as

$$\mu/e = 207 \cong 2\pi^4 = 195$$
$$\tau/\mu = 16.6 \cong 2^4 = 16$$

where

$$\mu = 0.106 \text{ GeV/c}^2$$
$$\tau = 1.777 \text{ GeV/c}^2$$

Based on these ratios we take the estimates in Figs. 18.1 and 18.2 and determine the second and third generation fermion masses using the fermion sequences of section 18.2. For the second generation we use the ν'/e ratio to find

$$\nu_2' = \nu'\mu/e = 1.7\times10^{-2} \text{ GeV/c}^2$$

Then we use the multipliers for each sequence to obtain the masses of the other fermions in the sequence.

The process determines the second and third neutrino masses from the ratios of the charged leptons. Then each sequence of masses is generated from the sequence's multiplier starting with the sequence's lepton mass.

First Generation

	SEQUENCE 1						SEQUENCE 2						
	e		u		c		t	ν'		d		s		b
Mass:	0.511×10^{-3}		1.80×10^{-3}		1.28		171	8.4×10^{-5}		4.24×10^{-3}		102×10^{-3}		4.34
Multiplier:		$2^5\pi$		$2^5\pi$		$2^5\pi$			32		32		32	

Sequence 1 multiplier: $2^5\pi = 100.5 \cong 100$;   Sequence 2 multiplier 32

Second Generation

Ratio: $207 \cong 2\pi^4 = 195$

μ		u_2		c_2		t_2	ν_2'		d_2		s_2		b_2
0.106		10.6		1060		10^5	1.7×10^{-2}		54×10^{-2}		1741		55706
	$2^5\pi$		$2^5\pi$		$2^5\pi$			32		32		32	

Figure 20.1. Projected approximate Second Generation fermion masses.

$\nu_e = 3.878 \times 10^{-11}$ GeV/c^2 = .03878 ev/c^2 consistent with known data estimates

$\nu' = \nu_e\,(\alpha_0{}^2)^{-1} = 0.03878$ ev/c^2 $\times 4096^2/e^2 = 8.4 \times 10^{-5}$ GeV/c^2 where e = 2.718.

$\nu_2' = \nu'\mu/e = 1.7\times10^{-2}$ GeV/c^2

Note: The physical neutrino mass is $\nu_2 = \nu_2'\,U(0)^2 = \nu_2'\,e^2/4096^2 = 7.5 \times 10^{-9}$ GeV/c^2 = 0.00075 ev/c^2

We now set $\nu_3' = \nu_2'\,\tau/\mu = 0.272$ GeV/c^2.

Third Generation

Ratio: $\tau/\mu = 16.6 \cong 2^4 = 16$

SEQUENCE 1							SEQUENCE 2						
τ		u_3		c_3		t_3	ν_3'		d_3		s_3		b_3
1.777		178		17800		1.78×10^6	0.272		8.7		279		8912
	100		100		100			32		32		32	

Note: The physical neutrino mass is $\nu_3 = \nu_3'\,U(0)^2 = \nu_3'\,e^2/4096^2 = 1.2\times10^{-7}$ GeV/c^2 = 1.2×10^4 ev/c^2

Figure 20.2. Projected approximate Third Generation fermion masses.

21. Higgs Mechanism and Cosmos Particle Masses

The Cosmos Theory sequence of fermion masses was based on an analysis of fundamental fermion masses. Higgs Models are also based on fits to experimental mass data. Therefore they cannot conflict because of their common origin in the fermion mass spectrum.

The accuracy of the Cosmos analysis masses, and the assumed accuracy of Higgs Model fits for masses, imply a relation between Higgs Model parameters and the Cosmos fits. Since they both have the same phenomenological source *there can be no conflict* in their respective results.

It is therefore a matter of taste as to which to select since both Cosmos masses and Higgs masses are phenomenologically equivalent. *A map between the Cosmos patterns of masses and Higgs Model parameters would be of interest.*

Cosmos Theory and its coupling constants and masses reflect an underlying theoretical framework that embodies the Reality that might be present in a "true" dynamic quantum field theory of space, matter, and energy.

The fundamental theoretical framework of Reality must be independent of space and time, and thus is not a familiar type of dynamical theory. It seems to be combinatoric in form since it is based on integer relations, and a few fundamental constants e, and π.

22. The Masses of Cosmos Universes

This chapter develops a condition for the determination of the initial mass-energy of a universe through an interaction or *ab initio*. A universe is viewed as a type of particle. A universe in its initial state has dimension r and resides in a parent universe of dimension r' > r. The contents of the universe initially have a high temperature T_r that causes its contents to consist of interaction-less, massless particles. We treat the contents of a universe as a perfect gas initially satisfying the Ideal Gas Law.

We assume the initial universe at the creation point is bound by a Casimir vacuum energy force that momentarily balances the pressure of its internal mass-energy contents.

This scenario gives a consistency condition similar to that in chapter 11. We assume the universe is initially an r dimension sphere in an r' dimension parent universe.[62] The calculation of the internal pressure is the same as in chapter 11.

The calculation of the Casimir force assumes an r' dimension space with a dimension array with $N_{r'} = 2^{r'+4}$ fermions. The estimates of the volume and surface area of the sphere are the same as in chapter 11.

We thus arrive at the modified (eq. 11.18) balance of pressure and Casimir force

$$0 = N_r \Gamma((r-1)/2+1)kT/(\pi^{(r-1)/2}a^{(r-1)}) - N_{r'}\pi^{(r-1)/2}a^{-r}/[r\Gamma((r-1)/2+1)] \qquad (22.1)$$

implying

$$E_r = N_r kT_r = N_{r'}\pi^{(r-1)}(ar)^{-1}/[\Gamma((r-1)/2+1)^2] \qquad (22.2)$$
$$= N_{r'}\pi^{r-1}(r/2e)^r)/(a\pi r^2)$$
$$= N_{r'}(2e\pi/r)^r/(a\pi^2r^2) \qquad (22.3)$$

after using Stirling's approximation for Γ functions (eq. 11.24). E_r is the total mass-energy for a universe of dimension r created in a universe of dimension r'.

22.1 Universe Particle Mass Spectrum

The universe mass spectrum depends on the dimension array size of the universe, within which it directly resides:

$$E_{r,r'} = 2^{r'+4}(2e\pi/r)^r/(a\pi^2r^2) \qquad (22.4)$$

relates the mass-energy $E_{r,r'}$ to r and r' and the initial radius a of the dimension r universe.

Eq. 22.4 shows a joint sequence of universe energy as a function of r and r'.

---

[62] The ultimate parent universe is the dimension 18 universe or a subuniverse of it.

For our r = 4 universe if it resides in an r' = 6 Megaverse we find

$$E_{4,6} = 2^6(e\pi/2)^4/(a\pi^2) = 2155/a$$

suggesting our universe in this case has an initial radius of

$$a = 2155/E_{4,6}$$

Using the total mass-energy of the observable universe $E_{4,6} = 3.49 \times 10^{82}$ GeV/c^2
We find the initial radius[63] (before/simultaneous with the Big Bang)

$$a = 6.17 \times 10^{-80}\ (\text{GeV/c}^2)^{-1}$$

Using the spatial meter – energy relation

$$m \equiv 0.89 \times 10^{12}\ (\text{GeV/c}^2)^{-1}$$

we find the initial radius a of our universe is

$$a = 5.55 \times 10^{-68}\ \text{m} \qquad (22.5)$$

22.2 Universe Temperature

The product of the initial temperature of a universe and its radius a satisfies

$$akT_r = N_{r'}/N_r\,(2e\pi/r)^r/(\pi^2r^2) = 2^{r'-r}(2e\pi/r)^r/(\pi^2r^2)$$

As $a \to 0$ and $kT \to \infty$, the product remains constant.[64]

22.3 Temperature Related to Mass as in the c_g Relation

Eq. 22.4 can be reexpressed as a temperature – mass relation:

$$kT_{r'} = 2^{r'-r}(2e\pi/r)^r/(\pi^2r^2)\ m \qquad (22.6)$$

by setting $m = a^{-1}$. This equation is analogous to the relation we found previously in the gambol analysis of particles:

$$kT = c_g m \qquad (22.7)$$

where[65]

$$c_g = 0.0785 \cong \pi^2/128 \qquad (22.8)$$

We define the quantity

$$C_g = 2^{r'-r}(2e\pi/r)^r/(\pi^2r^2) = \tfrac{1}{4}\,\pi^2\,(e/2)^4 = 8.41 \cong 0.852\pi^2$$

---

[63] Blaha (2004) has a very small but non-zero radius at the Big Bang point.

[64] This result is similar to those of Big Bang models such as this author's Blaha (2004).

[65] c_g appears often in internal particle gambol models.

for r = 4 and r' = 6 and find

$$c_g/C_g = 1/(.852 \text{ x } 128) = 1/109 = 0.0092$$

22.4 Temperature – Mass Relation kT = c_gM

The relation in eq. 22.6 was based on identifying $a^{-1} = m$. If we instead require

$$kT_{r'} = c_g\ m \tag{22.9}$$

then

$$2^{r'-r}(2e\pi/r)^r/(a\pi^2r^2) = ¼\ \pi^2\ (e/2)^4/a = \pi^2/128\ m$$

for r = 4 and r' = 6 and so

$$a = 32(e/2)^4/m = 109/m \tag{22.10}$$

Eq. 22.10 implies eq. 22.8 for c_g.

22.5 Expanded Energy Formula

We now reexpress the Energy expression of eq. 22.4 in terms of m using eq. 22.10.

$$\begin{aligned} E_{r,r'} &= 2^{r'+r+4}\ r^{-r-2}(e)^r\pi^{r-2}\ /a \\ &= 2^{r'+r+3}r^{-r-2}e^{r-4}\pi^{r-2}\ m \end{aligned} \tag{22.11}$$

Thus there is an analogy between universes and particles. Universe energies exhibit a joint sequence in r and r'.

23. Fermion Masses From Universe Mass Models

In the previous chapter we considered the initial stage of the creation of universes. We saw that the mass-energy of a universe emerged from the initial balance of internal mass-energy pressure and external Casimir force.

We now turn to consider a similar scenario for fundamental particles.[66] We assume an elementary particle has an mass of similar form to a universe within a parent universe given by eq. 22.12. In the case of our universe, the universe external to the particle has dimension r' = 4 and the particle has internal dimension r.

$$M_{r,r'} = 2^{r'+r+3} r^{-r-2} e^{r-4} \pi^{r-2}\, m \qquad (23.1)$$
$$= 2^{r+7} r^{-r-2}\, e^{r-4}\, \pi^{r-2} m$$

where m is a mass parameter common to all eight fundamental fermions.

There are two sequences of fermions:

Sequence 1: ν' d s b
Sequence 2: e u c t

Section 18.2 gives the sequences:

<u>Sequence 1</u>

1 $\nu' \cong \pi e 2^{-15}$ GeV/c^2/π
1 $d = \pi e 2^{-11}$ GeV/c^2
¾ $s = ¾ \times \pi e 2^{-6}$ GeV/c^2
½ $b = \pi e 2^{-1}$ GeV/c^2

<u>Sequence 2</u>

1 $e = \pi e 2^{-14}$ GeV/c^2
¾ $u = ¾ \times \pi e 2^{-12}$ GeV/c^2
¾ $c = ¾ \times \pi^2 e 2^{-4}$ GeV/c^2
½ $t = 2\pi^3 e$ GeV/c^2

The above sequence mass values can be interpreted as reflecting a combination of an inner *universe shell* within each fermion and a "bare" fermion mass. We now consider this model for each fermion.

We view the sequence 1 fermions as having an internal universe shell for values of r using eq. 23.1 modified to

[66] We considered similar possibilities in Blaha (2021d), (2022d) and (2023e).

$$M_{r,r'i} = 2^{r+7} r^{-r-2} e^{r-4} \pi^{r-2} m_{0i} = c_{Shellr}\, m_{0i} \quad (23.2)$$

with

$$c_{Shellr} = 2^{r+7} r^{-r-2} e^{r-4} \pi^{r-2} \quad (23.3)$$

where i labels the fermion and m_{0i} is the bare mass.

Sequence 1 Fermion Masses (23.4)

r = 3 gives $M_{3,4\nu'} = 2^{10} 3^{-5} e^{-1} \pi\, m_{0\nu'} = c_{Shell3}\, m_{0\nu'}$
Since $\nu' \cong e2^{-15}$ GeV/c^2 we find $m_{0\nu'} = \nu'/c_{Shell3}$

r = 3 gives $M_{3,4d} = 2^{10} 3^{-5} e^{-1} \pi\, m_{0d} = c_{Shell3} m_{0\nu'}$
Since $d = \pi e 2^{-11}$ GeV/c^2 we find $m_{0d} = d/c_{Shell3}$

r = 3 gives $M_{3,4s} = 2^{10} 3^{-5} e^{-1} \pi\, m_{0s} = c_{Shell3}\, m_{0s}$
Similarly we find $m_{0d} = = s/c_{Shell3}$

r = 3 gives $M_{3,4b} = 2^{10} 3^{-5} e^{-1} \pi\, m_{0b} = c_{Shell3}\, m_{0b}$
Similarly we find $m_{0b} = = b/c_{Shell3}$

These four sequence 1 fermions have an inner shell of 3 dimensions. Note fermion mass is denoted by its "letter."

We view fermion sequence 2 as also having internal universe shells for values of r. This sequence indicates an internal universe shell structure based on eq. 23.1.

Sequence 2 Fermion Masses[67] (23.5)

r = 5 gives $M_{5,4t} = 2^{12} 5^{-7} e \pi^3 m_{0t} = c_{Shell5}\, m_{0t}$
$m_{0t} = t/c_{Shell5}$

r = 4 gives $M_{4,4c} = 2^{11} 4^{-6} \pi^2 m_{0c} = c_{Shell4}\, m_{0c}$
$m_{0c} = c/c_{Shell4}$

r = 3 gives $M_{3,4u} = 2^{10} 3^{-5} e^{-1} \pi\, m_{0u} = c_{Shell3}\, m_{0u}$
$m_{0u} = u/c_{Shell3}$

r = 2 gives $M_{2,4e} = 2^{9} 2^{-4} e^{-2} m_{0e} = c_{Shell2}\, m_{0e}$
$m_{0e} = c/c_{Shell2}$

There appears to be reason to view fermions as having an internal universe shell based on the variations in powers of π and the base of natural logarithms e in factors of π/e. *This π variation based on universe shell dimension provides an "explanation" of the*

[67] It is possible to have higher dimension "bubbles" within lower dimension universes through a judicious choice of boundary conditions.

appearance of e and π in coupling constants and masses due to universe-particle consistency conditions.

24. Geometric Reasons for the Thread Sequences: of Spaces, Coupling Constants, Gravitation, and Masses

Cosmos Theory has sequences of spaces of which 10 spaces are Physical in the sense of supporting universes. We have found sequences of Coupling Constants, of up-type fermion masses, of down-type fermion masses and of universe masses.

These threads of Reality raise the questions of their origin and their forms. We now consider the origin of each type of sequence.

24.1 The Sequence (Spectrum) of Cosmos Spaces

The sequence of Cosmos spaces is the consequence of the number of totally antisymmetric tensors of dimension r and the number of creation and annihilation operators in a Fourier expansion of a Second Quantized Fermion wave function. To avoid duplications dimensions are constrained to be even numbers. The Physical sequence is limited by a consistency condition described earlier in this book. See chapters 1 – 10 for more detail.

24.2 The Sequence of Coupling Constants

We have shown coupling constants appear to form a powers sequence as seen earlier. We now consider the expression of a coupling constant as the product of a number of factors that are based on geometry.

The number of fermions in a fundamental representation of SU(n) is n. The SU(n) coupling constant g(n) has the form:

$$\begin{aligned} g(n) &= (4\pi)^{1/2}\, e2^{n-6} = e\,(2^{-10}\pi)^{1/2}\, 2^n = 0.1505\; 2^n \\ &= g_g\,(\text{number of spin states per fermion})^{\text{number of fermions}} \end{aligned} \tag{24.1}$$

The coupling constant is a fundamental value times the number of spin states (a geometric property of fermions equal to half the column length of a spinor) raised to the number of fermions. Thus it is a product of degrees of freedom.

We see that g(n) has n factors – one factor for 2 spin states for each fermion in the SU(n) fundamental representation. The origin of the constant in e and π:

$$g_g = e\, 2^{-5}\pi^{1/2} \tag{24.2}$$

appears to be due to the multi-dimensional geometry volume and surface values according to the previous results. These views follow from the consideration of fermion masses.

The factor of $2^{-5} = 1/32$ appears to result from the number of Normal fermions, 32, in each dimension 4 UST dimension array of fermions. An analysis of fermion masses supports this view.

The two sequences of fermion masses are

	SEQUENCE 1 “up-type”							SEQUENCE 2 “down-type”						
	e		u		c		t	ν'		d		s		b
Mass:	0.511×10^{-3}		1.80×10^{-3}		1.28		171	8.4×10^{-5}		4.24×10^{-3}		102×10^{-3}		4.34
Multiplier:		$2^5\pi$		$2^5\pi$		$2^5\pi$			32		32		32	

They implement the multipliers:

Up type fermion sequence multiplier: $2^5\pi = 32\pi$
Down type fermion sequence multiplier: $2^5 = 32$

The multiplier of $2^5 = 32$ appearing in both sequences is the number of Normal (non-Dark) fermions in a UST layer. It is based on the concept that all fermions of a layer are intertwined.

The 2^{-5} factor for coupling constants “combines” with the 2^5 multiplier for fermion masses to form a cancellation. The factors 1/32 and 32 result from the number of Normal fermions in one UST layer. (UST has four layers of fermions.)

An important point is the number of factors of 2 of the SU(n) coupling constant is equal to the number of fermions in the SU(n)’s fundamental representation. *The coupling constant “felt” by one fermion depends on the total number of fermions in the group’s fundamental representation.*

The coupling constant expressions follow from geometry and the count of the number of fundamental fermions in each symmetry group representation.

24.3 The Sequences of Fermion Masses

There are two sequences of fermion masses. The growth of the sequences is tied to the universe shells within the fermions. Eqs. 23.4 and 23.5 show the particle masses are a product of a universe shell factor (eq. 23.2) and a “bare” mass value. The multiplier factors appear directly. The appearance of e and π is explained as due to volume and surface factors in the fermion mass model based on the universe consistency condition applied for fermion stability.

The form of the e and π entries is the reason for viewing fermions as having internal universe shells based on the variations in the powers of π and e, the base of natural logarithms. *This π and e variations, based on internal universe shell dimension, provide an “explanation” of the appearance of e and π in coupling constants and masses due to volume and surface factors in the universe-particle consistency conditions.*

In the type 2 sequence we see the masses vary as:

$$v', \quad d = 32v', \quad s = 32d, \quad b = 32s$$

Consequently the masses have $\mathbf{2^{5n} = 32^n}$ ratios where n = 1, 2, 3. Sequence 2 fermion masses increase by factors of 32 just like coupling constants increase by factors of 2. *The coupling constant increase factor of 2 reflects the SU(2) of fermion spins. 2. The masses increase factor of 32 reflects the (broken) SU(32) symmetry group of each UST layer of fermions.*

In the type 1 sequence we see

$$e = 2\pi v', \quad u = 32\pi e, \quad c = 32\pi u, \quad t = 32\pi c$$

Consequently the masses have a $\mathbf{2^5\pi/e}$ multiplier leading to a $(2^5\pi/e)^n$ due to internal universe shells based sequence of masses by eq. 23.5 and Fig. 3.2. (Note the masses for the lower fermions in the sequence show some deviation from their experimental values.).

The powers of $2^5 = 32$ in both sequences reflect the number of Normal fermions in each UST layer.

Note: the masses for the lower fermions show some deviation from the experimental values. We have chosen to approximate in order to display the overall pattern.

The fermion masses are governed by geometry and the number of Normal fermions (degrees of freedom) in a UST layer.

24.4 The Sequence of Universe Mass-Energies

The sequence of universe mass-energies $E_{r,r'}$ is based on a universe creation consistency condition which is in turn based on geometry and the quantum vacuum:

$$E_{r,r'} = 2^{r'+4}(2e\pi/r)^r/(a\pi^2r^2) \qquad (22.4)$$

for a universe of dimension r, and radius a, created within a universe of dimension r'.

Eq. 22.4 is based on a consistency condition for the creation of a universe where the pressure in the universe equals the Casimir force generated by the enclosing universe. The mass-energy $E_{r,r'}$ shows a joint sequence in powers of 2 of the enclosing universe's dimension and powers of $(2e\pi/r)$ of the universe's dimension.

The basis of the form of eq. 22.4 is the forms of surface area and volume in various dimensions, and of Cosmos Theory dimension array sizes (numbers of fermions).

These results suggest that a universe is a form of particle.

24.4.1 Gravitation G_r in r Dimensions

The gravitation coupling constant G_r in r dimensions is associated with the symmetry group $SU(n) = SU(2^{r+4})$:

$$G_r = g_r^2/4\pi \, M_r^{-2}$$

with

$$g_r^2/4\pi = e^2 2^{2n-12}$$

where $n = 2^{r+4}$ and M_r is a mass.

The mass factor M_r^{-2} is required on dimensional grounds in gravitational dynamic equations.[68] This factor causes gravitation G to be weak.

We appear to have a viable estimate of the value of G and its origin in a coupling constant formulation.

24.5 Unifying Principle for The Sequences

The seven sequences that appear: Cosmos spaces, Coupling Constants, Gravitation coupling constant, Up-Type fermions, Down-Type fermions, Vector Boson W and Z masses, and universe masses are all based on Mathematics and Geometry:

1. Cosmos spaces based on Mathematics of totally asymmetric tensors (geometry of surfaces and volumes) and fermion creation and annihilation operators.

2. Coupling Constants based geometry and spin of fermion creation and annihilation operators.

3. Masses based on geometry and the Cosmos symmetry group of the UST layer.

4. Universe masses based on geometry.

5. Vector boson: W and Z masses based on coupling constants.

We conclude the total Lagrangian for our universe is fixed:

3. Cosmos Theory gives spaces and dimension arrays, coupling constants and masses.

4. The UST Lagrangian is almost totally defined by using coupling constants and masses (and scalar and vector bosons).

The constants in the total elementary particle Lagrangian is thus known except for the role of Higgs bosons. Possible additional particles and interactions remain to be found.

The above sequences of spaces, coupling constants, masses, and universe masses all reflect a fundamental basis in the features of the range of dimensions in Geometry from totally asymmetric tensors through particles with internal universe shells. A fait Accompli!

[6868] Models which change gravity from 1/r Newtonian gravity usually change the space-time dependence – not G.

The Unified SuperStandard Theory (UST)

25. The Unified SuperStandard Theory (UST) of Our Universe

The Unified SuperStandard Theory is important for two reasons: its dimension array form is from the $n_H = 3$ Hilbert-like construction order. It provides experimental evidence for the construction pattern of Cosmos spaces. It also provides a direct generalization of the Standard Model with demonstrable reasonable extensions.

The pattern of the experimentally known three generations of fermions and the known Standard Model groups of Figs. 25.1 – 25.3 demonstrate the validity of the known part of the n = 3 Cosmos Theory dimension array.

The form of the Unified SuperStandard Theory (UST) that this author developed some years ago was shown by the author to follow from the hypercomplex Cayley-Dickson number n = 3 space part of HyperCosmos spaces in the Fall of 2019. Embolden by this remarkable result the author proceeded to develop Cosmos Theory in the subsequent years. In this chapter we outline some features of UST in relation to Cosmos Theory.[69] One of the remarkable results is the necessity of the author's Generation and Layer groups that he originally based on partially conserved baryon numbers.[70]

The chapter also provides a brief outline of the n = 4 Cosmos space that describes our Megaverse model with 6 space-time dimensions.

25.1 The n = 3 UST Features

The initial form of UST was based on The Standard Model. It was described in our books of 2017 and 2018 as well as in earlier books. Fig. 25.2 shows the known parts of the Standard Model fundamental fermions within the UST extension. The groups of the Standard Model within the UST are displayed in the first layer in Fig. 25.3 and 25.4.

The UST is described in Blaha (2023d) as well as in earlier books by the author with parts dating back to Blaha (2018e) though (2002) where the Two Tier Renormalization formalism is first described. *This formalism applies to renormalization in any space-time dimension and thus makes quantum field theories in higher dimensions possible.*

25.1.1 Extension to Generation, Layer and Connection groups.[71]

The original introduction of the Generation and Layer groups, which motivated the extension of the fermion spectrum to four layers[72] with each containing four

---

[69] The figures in this chapter were presented in Blaha (2020d), (2020e), and (2020i) as well as other books by the author. Also see Blaha (2023d) for more detail.

[70] See the original presentations of these groups in chapters 12 and 13 of Blaha (2018e), and also Blaha (2017b) and (2017c).

[71] See appendices 25-A and 25-B here, or 9-A and 9-B of Blaha (2023d) for a detailed discussion.

[72] The introduction of the Generation and Layer groups and the consequent expansion of the fermion sector to four layers of four generations implied the Cabibbo-kobayashi-Maskawa matrix (CKM) should be correspondingly

generations, was done by the author in 2017 based on a consideration of broken baryon number symmetries. The Connection groups were introduced later to interconnect the normal and dark fermion sectors and further connect the four layers. Together these enhancements yield 256 fundamental fermions[73] and a total of 256 symmetry group and space-time dimensions. The remarkable match of this number of dimensions and the size of the n = 3 dimension array in 2019 led the author into hypercomplex number based Cosmos spaces.

25.1.2 UST Structure

A comparison of the $n = n_H = 3$ case and the structure of fermions and symmetry groups shows that the structure of UST is firmly based on Cosmos Theory n = 3 space.[74]

The features of the UST are described in detail in numerous books by the author including Blaha (2023d) and books listed in the footnotes in this chapter and in References later. Clearly the n = 3 Cosmos space correctly describes the UST structure and thus the known Standard Model.

25.1.3 UST Interactions

Dirac equations for fermions; and Yang-Mills equations for vector bosons with the details of the equations are determined by the relevant symmetry groups. General Relativistic generalizations of the equations are immediate. Higgs Scalar boson particles are also determined. See Blaha (2022f) or Blaha (2018e) for the PseudoQuantum formulation of the Riemann-Christoffel curvature tensor whose terms lead to the Euler-Lagrange formulation of the dynamical equations.

25.2 The n = 4 Megaverse Features

The ten Cosmos spaces support a universe of the type of the n = 4 Cosmos space. This universe might be viewed as the parent universe of our universe. The Megaverse structure may be viewed as composed of four copies of our universe's structure as Fig. 25.4 shows.

The Megaverse has 1024 fundamental fermions. They may be viewed as four copies of our universe's fermions. See Fig. 25.5. The Megaverse has groups that are four copies of our universe's groups that are modified to have Connection groups as appendix 25-B describes.

---

expanded in principle from a 3 × 3 to a 16 × 16 matrix. The presumed weakness of the Layer group interactions would allow it to be approximated by a 4 × 4 matrix.

[73] As a matter of principle the author chose "equality" between the Normal and Dark sectors realizing the connection between them would be very weak interactions.

[74] It is possible to combine symmetries in forms that mask the structure of the UST.

	Normal		Dark	
	4	4	4	4

Layer 1

Generation				
1	e q-up	v q-down	e q-up	v q-down
2	μ q-up	v q-down	μ q-up	v q-down
3	τ q-up	v q-down	τ q-up	v q-down
4	e′ q-up	v q-down	e′ q-up	v q-down

Layer 2

Generation				
1	e q-up	v q-down	e q-up	v q-down
2	μ q-up	v q-down	μ q-up	v q-down
3	τ q-up	v q-down	τ q-up	v q-down
4	e′ q-up	v q-down	e′ q-up	v q-down

Layer 3

Generation				
1	e q-up	v q-down	e q-up	v q-down
2	μ q-up	v q-down	μ q-up	v q-down
3	τ q-up	v q-down	τ q-up	v q-down
4	e′ q-up	v q-down	e′ q-up	v q-down

Layer 4

Generation				
1	e q-up	v q-down	e q-up	v q-down
2	μ q-up	v q-down	μ q-up	v q-down
3	τ q-up	v q-down	τ q-up	v q-down
4	e′ q-up	v q-down	e′ q-up	v q-down

Figure 25.1. Periodic table for the Normal and Dark fermions of UST. Each fermion corresponds to a dimension. Subscripts of neutrinos and quarks are not shown. A fourth generation is indicated for each layer. There are 256 fermions corresponding to the 256 entries in the dimension array for n = 3. This figure is from Blaha (2020d) for the UST.

The n = 3 UST Fermion Periodic Table

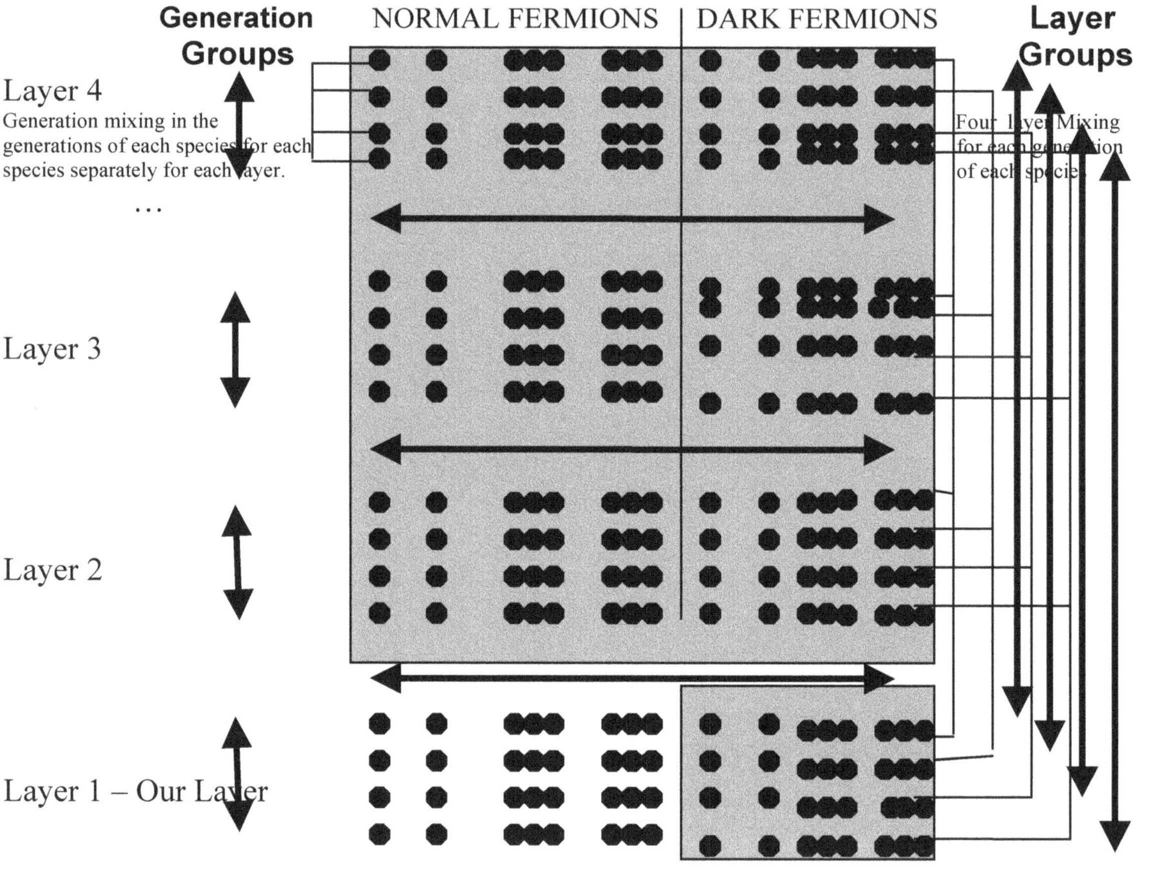

Figure 25.2. Fermion particle spectrum and partial example of pattern of mass mixing of the Generation and Layer grroups. Unshaded parts are the known fermions including an additional, as yet not found, 4th generation shown. The lines on the left side (only shown for one layer) display the Generation mixing within each layer's species. The Generation mixing applies within each layer using a separate Generation group for each layer. The lines on the right side show Layer group mixing with the mixing amongst all four layers for each of the four generations individually. There are four Layer groups. For each generation and each layer, SU(2)⊗U(1) mixes between an e-type fermion and a neutrino-type fermion. It also mixes between an up-quark-type fermion and a down-quark-type fermion. SU(3) mixes among each up-quark triplet and down-quark triplet separately. Complex Lorentz group transformations map among all four fermions: Dirac ↔ tachyon ↔ up-quark ↔ down-quark There are 256 fundamental fermions counting quarks as triplets. This figure is from Blaha (2020e) for the UST.

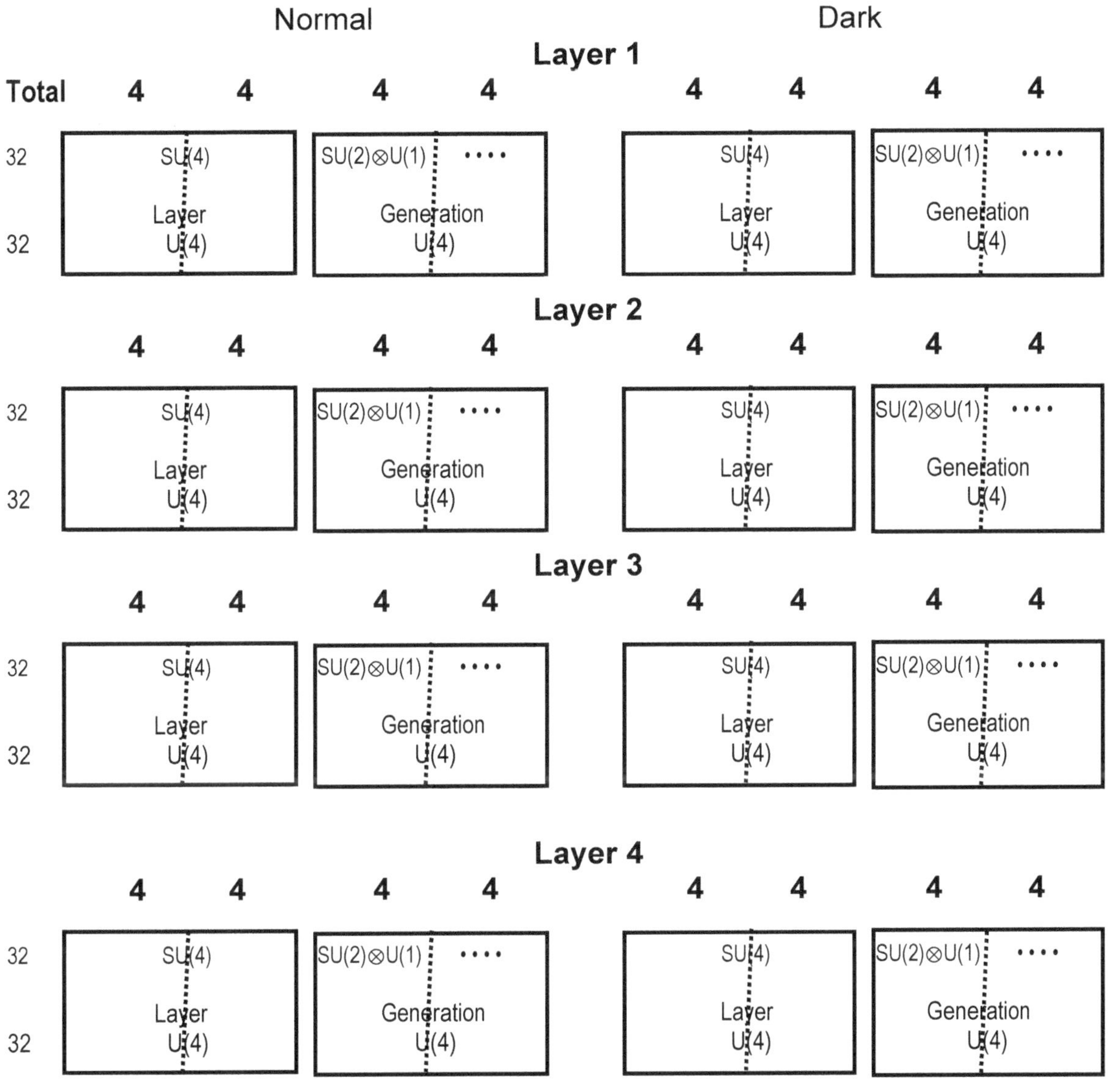

Figure 25.3. Schematic for the Normal and Dark symmetries of UST. SL(2, C) Groups diifer in different layers. The real and imaginary dimensions are counted separately. For example SU(4) has four real dimensions and four imaginary dimensions. Each layer has 32 + 32 = 64 dimensions. There are 256 dimensions corresponding to the 256 entries in the dimension array for UST where n = 3. The 32 • 's form sets of 4 dimensions allocated initially to 4D space-times. These dots are mapped to 7 SU(2)⊗U(1) Connection groups in Fig. 25.4. Three sets of dots unite layers, and 3 sets of dots unite corresponding Normal and Dark layers This figure is from Blaha (2020d) for the UST.

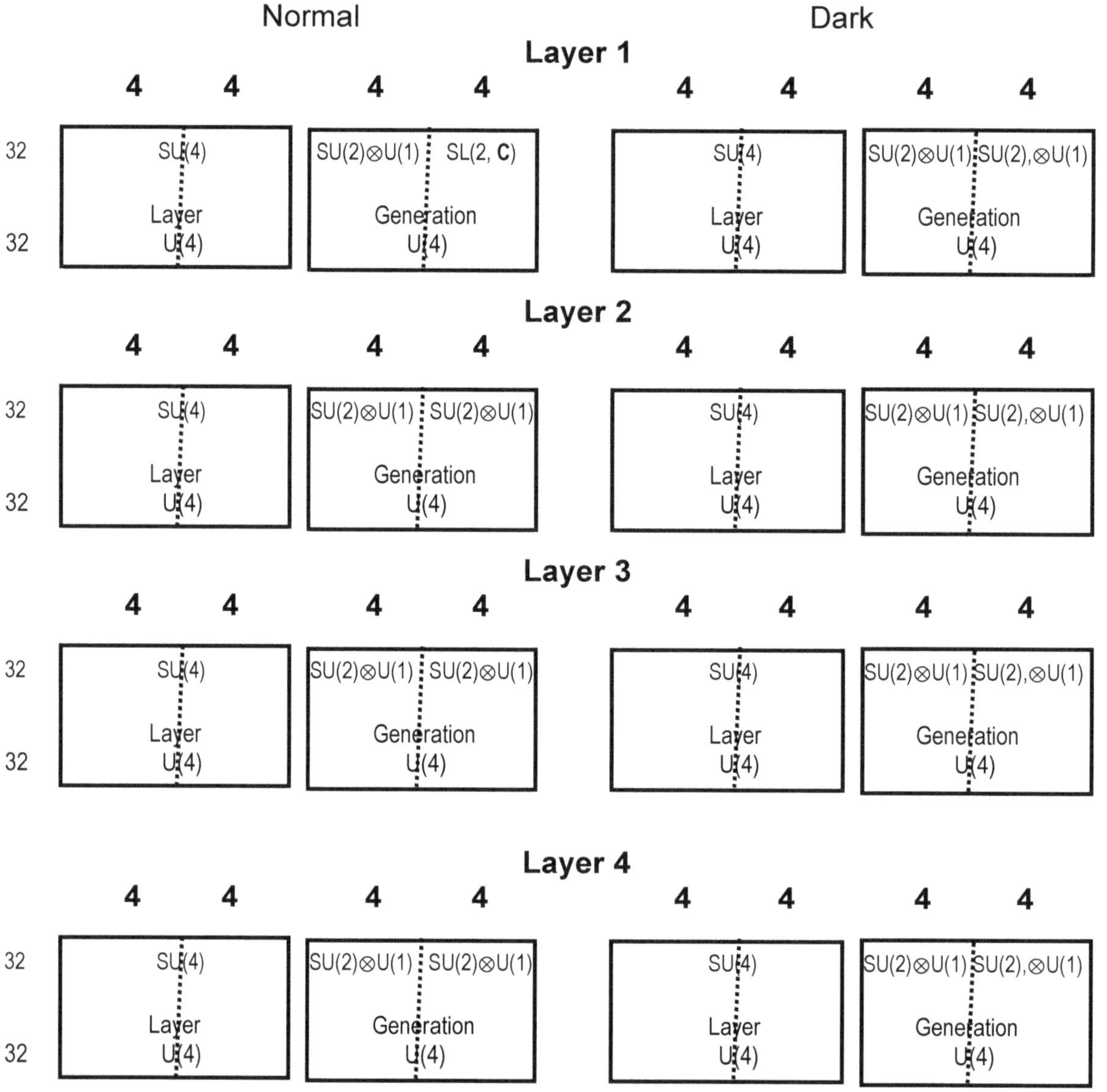

Figure 25.4. Schematic for the Normal and Dark symmetries of UST. The sets of dots in Fig. 25.3 are replaced by SU(2)⊗U(1) Connection groups. The SL(2, C) represents the Lorentz group $SO^+(1,3)$. This figure is from Blaha (2020d) for the UST.

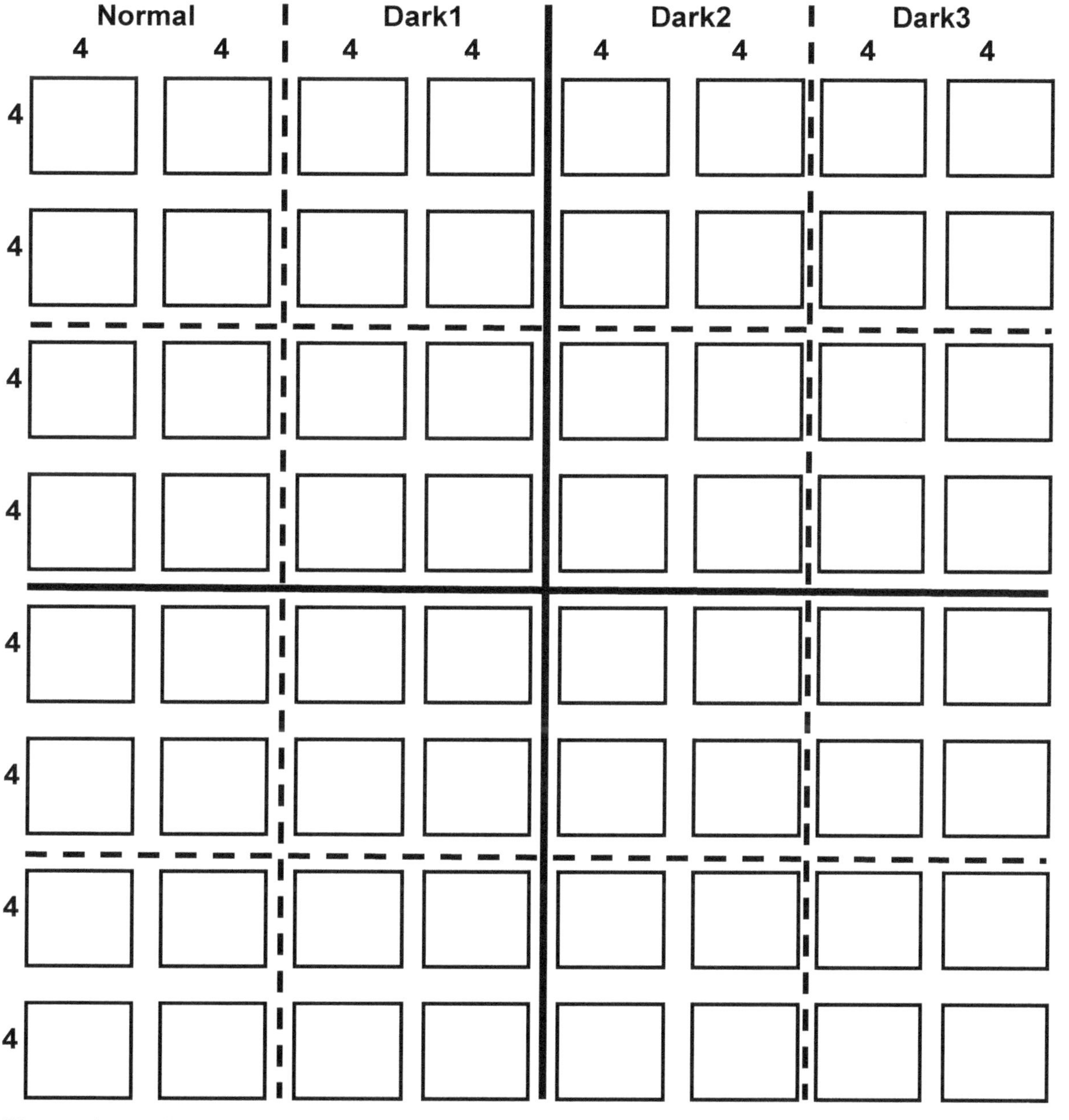

Figure 25.5. The n = 4 Megaverse dimension array. Four quadruplings are visible. Each quadrant is an n = 3 UST dimension array. There are sixty-four 4 × 4 n = 1 blocks. Each 8 × 8 quadrant is an n = 2 dimension array. Each 16 × 16 quadrant is an n = 3 dimension array. An exact match with Hilbert orders is evident. This figure is from Blaha (2020d) for the UST.

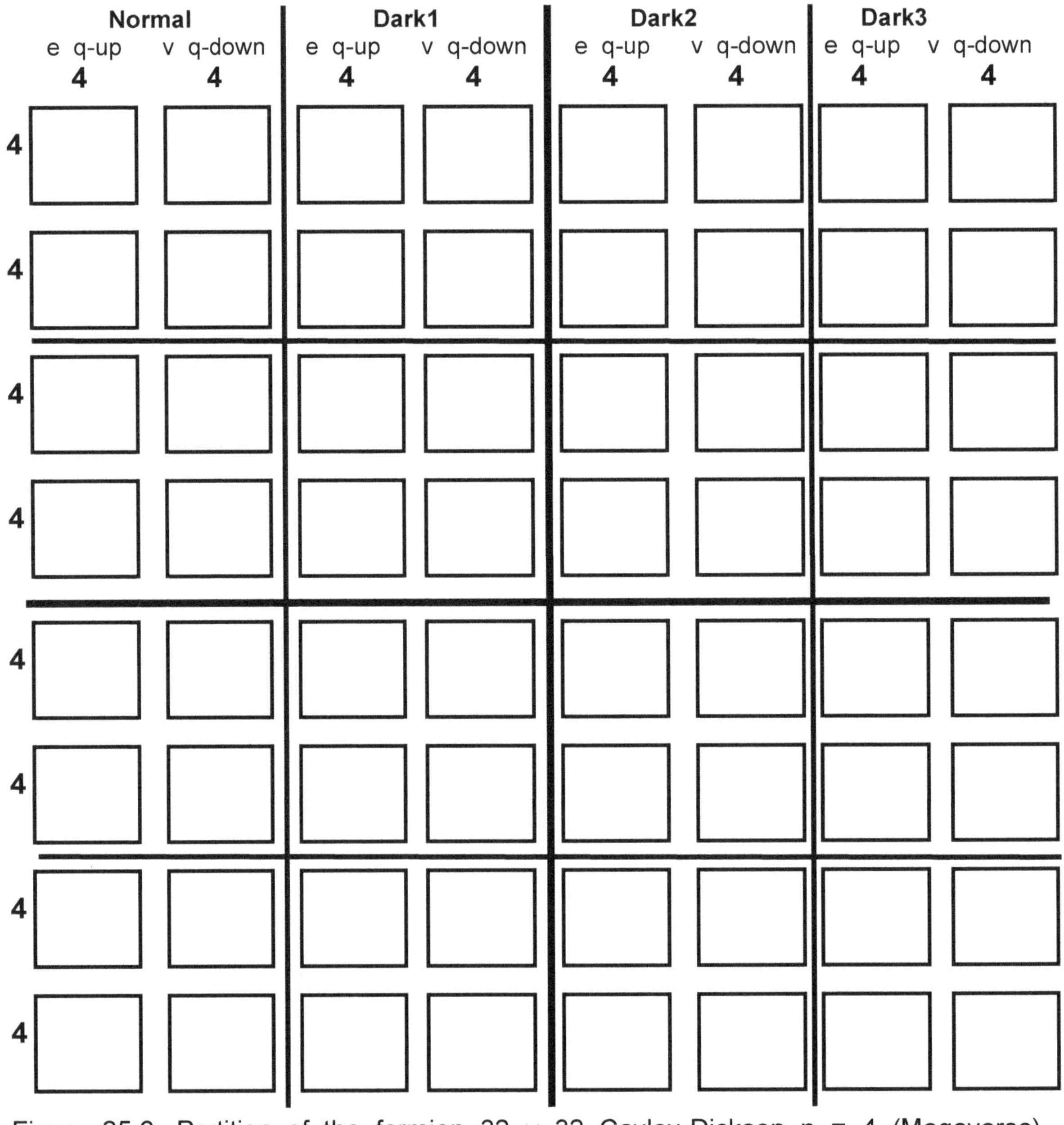

Figure 25.6. Partition of the fermion 32 × 32 Cayley-Dickson n = 4 (Megaverse) dimension array. The label e q-up indicates a charged lepton – up-type quark pair. The label v q-down indicates a neutral lepton – down-type quark pair, and so on. This figure is from Blaha (2020i) for the UST.

Appendix 25-A. Generation and Layer Groups[75]

25-A.1 U(4) Generation Groups

In the Big Bang all particles were massless and all symmetries unbroken. Hence the four Normal particle number symmetries, and the four Dark particle number symmetries, are all "conserved" in the Big Bang. Afterwards conservation laws are then broken.

We define two particle number operators for normal up-quark particles and down-quark particles, B_{uq} and B_{dq}. Similarly we define two particle number operators for normal species "e" (electron) particles and species "ν" particles, B_e and B_ν. Similarly we define Dark matter equivalents:[76] B_{De}, $B_{D\nu}$, B_{Duq}, and B_{Ddq}.

In the absence of symmetry breaking these fermion particle number operators would be conserved. Thus there are two sets of "diagonal" operators with associated U(4) groups for the Normal and Dark sectors. They are part of the Normal U(4) Generation Group and the Dark U(4) Generation Group.

The fermion fundamental representation of a U(4) group has four fermions. U(4) has rotations, and also interactions of the form $\overline{\Psi}\gamma\cdot B\cdot T\Psi$ where Ψ is a fermion four-vector, B is a 16 component U(4) gauge field, and T consists of 16 component 4 × 4 U(4) arrays.

In the case of the Generation Group the gauge fields have electric charge zero. Since the four species have different electric charges (1, 0, 2/3, -1/3) the U(4) gauge boson fields cannot mix the fermions of different species. Generation Group interactions are diagonal[77] in fermion species (e, ν, up-quark, and down-quark species).

Consequently the U(4) Generation Group for each layer[78] must have a reducible representation D consisting of a set of four fundamental U(4) representations, D_e, D_ν, D_{upq}, and D_{dnq}, appearing in blocks along the diagonal of D. Each block is a separate U(4) irreducible representation for a species (due to the electric charge superselection rule.) There is a U(4) Generation group for each of the four layers of the Normal and Dark sectors totaling to 8 generation groups.

There are four generations of each species in the Normal and in the Dark matter sectors. The four generations for each fermion species: e, ν, up-quark, and down-quark each furnish a U(4) fundamental representation within the reducible representation D. The fourth generation of normal fermions has not as yet been found due to their extremely large masses.

---

[75] From Blaha (2019g), (2018e) and (2023d).

[76] By analogy, we assume that there are four species of Dark matter: charged Dark leptons, neutral Dark leptons, Dark up-type quarks, and Dark down-type quarks. Thus we are led to the Dark particle numbers: Dark Baryon Numbers, and Dark Lepton Numbers shown above.

[77] ElectroWeak interactions can cross between species due to their charged gauge vector bosons.

[78] For the Normal and Dark sectors separately.

The Generation Group rotates the fundamental fermions of each fundamental representation separately for each of the four species of each of the four layers.[79] Thus the Generation Group guarantees that all generations of each species have the same electric charge and other quantum numbers.

The U(4) Generation Group also specifies a gauge field interaction among the fermions of its fundamental representation, species by species, for both Normal and Dark sectors. The form of the interactions for the Normal sector for each fermion layer is:

$$g_e\overline{\Psi}_e\gamma\cdot B_e\cdot T\Psi_e + g_v\overline{\Psi}_v\gamma\cdot B_v\cdot T\Psi_v + g_{upq}\overline{\Psi}_{upq}\gamma\cdot B_{upq}\cdot T\Psi_{upq} + g_{dnq}\overline{\Psi}_{dnq}\gamma\cdot B_{dnq}\cdot T\Psi_{dnq} \tag{25-A.1}$$

where g_e … are coupling constants, *the* gauge vector fields are B_e … , and the Ψ_e … are 4-vectors of fermions of the four generations of each species in a layer.

The gauge vector bosons of the Generation Group have large masses. If the conservation of the fermion particle numbers is broken then we view it as a consequence of Generation Group symmetry breaking.

Generation Group rotations guarantee the internal quantum numbers of each generation of each species are the same since symmetry breakdown is not present at the instant of the Big Bang.

The above discussion applies similarly to the Dark sector. Thus there are 8 Generation Groups in total.

See Blaha (2019g) and (2018e) for a detailed discussion of the Generation Groups.

25-A.2 U(4) Layer Groups

The set[80] of particle number operators can be extended if we take account of the fourfold fermion generations.

We can subdivide the above particle number sets into four additional particle numbers *per generation*. For the ith generation (of the four generations) we define

L_{ie} – The "e" species particle number for the ith generation
L_{iv} – The ν species particle number for the ith generation
L_{iuq} – The up-quark species particle number for the ith generation
L_{idq} – The down-quark species particle number for the ith generation

L_{iDe} – The Dark "e" species particle number for the ith generation
L_{iDv} – The Dark ν species particle number for the ith generation
L_{iDuq} – The Dark up-quark species particle number for the ith generation
L_{iDdq} – Dark down-quark species particle number for the ith generation

[79] There are separate Generation groups for each layer.
[80] Here again, in the Big Bang all particles were massless and all symmetries unbroken. Hence particle numbers are "conserved" in the Big Bang. Conservation is then broken afterwards in most cases.

for each generation i = 1, 2, 3, 4. Individual fermions have positive $L_{ia} = +1$ values and antifermions have negative $L_{ia} = -1$ values for each species.

At this point we have a set of four particle number operators for each of four generations (i = 1, 2, 3, 4) of fermions in the Normal sector and similarly in the Dark sector. We then define a U(4) group framework for each set of particle numbers.

The only way to specify fundamental representations for each of the four sets in a sector is to assume there are four layers, with each layer having four generations, and with a fundamental U(4) representation defined for each generation composed of fermions from each layer. Thus there are four Layer Groups for each Normal and each Dark sector: a Layer Group for generation 1, a Layer Group for generation 2, and so on.

The Layer Groups are also "split" by species due to the electric charge superselection rule. Each Layer Group is diagonal in the four fermion species. All their gauge fields are electrically neutral. Thus the Layer Group fundamental representations of the Normal sector total 16.[81] The Layer Group fundamental representations of the Dark sector also total 16. There are 4 Normal Layer Groups and 4 Dark Layer Groups.

Consequently each of the four U(4) Layer Groups in the Normal fermion sector has a reducible U(4) representation D_j for j = 1, 2, 3, 4. Each reducible representation is composed of four irreducible U(4) representations for each species due to the electric charge superselection rule:

$$D_j = D_{je} + D_{jv} + D_{jupq} + D_{jdnq},$$

for j = 1, 2, 3, 4.

There are four layers of each species in the Normal and in the Dark matter sectors. The second, third and fourth layers of normal fermions has not as yet been found due to their extremely large masses.

A Layer Group rotates the fundamental fermions of each fundamental representation separately for each of the four species of each of the four generations.

The Layer Groups guarantee that all layers of each species have the same electric charge and other quantum numbers.

Each U(4) Layer Group also specifies a gauge field interaction among the fermions of its fundamental representation, species by species, for both Normal and Dark sectors. The form of the interactions is:

$$g_{ei}\overline{\Psi}_{ei}\gamma\cdot C_{ei}\cdot T\Psi_{ei} + g_{vi}\overline{\Psi}_{vi}\gamma\cdot C_{vi}\cdot T\Psi_{vi} + g_{upqi}\overline{\Psi}_{upqi}\gamma\cdot C_{upqi}\cdot T\Psi_{upqi} + g_{dnqi}\overline{\Psi}_{dnqi}\gamma\cdot C_{dnqi}\cdot T\Psi_{dnqi} \tag{25-A.2}$$

for i = generation = 1, … , 4, where g_{ei} … are coupling constants, the gauge fields are C_{ei} … , and the Ψ_{ei} … are 4-vectors of fermions formed of the ith generation fermions in each layer of each species.

---

[81] Four Layers groups irreducible representations for the generations × four species = 16 irreducible representations.

The gauge vector bosons of the Layer Groups also have large masses. If the conservation of the fermion particle numbers is broken then we view it as a consequence of Layer Groups symmetry breaking.

Layer Group rotations guarantee the internal quantum numbers of each layer of each species are the same since symmetry breakdown is not present at the instant of the Big Bang.

The above discussion applies similarly to the Dark sector. There are 16 Dark Layer Groups.

Fig. 25.2 shows the fundamental fermion spectrum with the representations of the Generation groups and Layer groups indicated.

Experimentally, we know of three generations of fermions—the lowest 3 generations of the lowest level. The remaining 4th generation and three layers of fermions are of much higher mass and are yet to be found.

See Blaha (2019g) and (2018e) for a detailed discussion of the Layer Groups. We note in passing that the symmetries of these number operators are badly broken. Yet the underlying group structure remains.

Appendix 25-B. Connection Group Symmetries and HyperCosmos Space-Time Coordinates

Hypercomplex numbers are known to be related to symmetry groups. In this section we consider the multiple space-time symmetries that appear in the separation of dimension arrays into representations of groups. We have suggested in Blaha (2021b), (2021e) and (2021g) that the set of hypercomplex space-time dimensions be transformed to a set symmetry groups in each HyperCosmos space.

There is a two-fold justification for this procedure: 1). There is no evidence for the existence of hypercomplex numbers in our universe. 2) The generated set of symmetry groups has the important purpose of providing ultra-weak interactions uniting Normal and Dark matter. Without unification, Dark matter becomes physically irrelevant for elementary particle physics.

We begin by noting the space-time dimension is set by the dimension array:

$$r = \log_2 (d_{dN}/16) \tag{25-B.1}$$

For N = 9 the space-time dimension is 0 since $d_{dN} = 16$. For N = 7 (our universe's space) the space-time dimension is 4 since $d_{dN} = 256$. We will consider the cases of QUeST, UTMOST, and Maxiverse to illustrate the method.

25-B.1 Hypercomplex Coordinates Transformed to Symmetry Groups in Our Universe N = 7

Our QUeST formulation for our N = 7 universe has a $d_{d7} = 256$ component dimension array. A preliminary view of the symmetry groups that it implies appears in Fig. 25-B.1. Note that there are 32 dimensions that are initially allocated to space-time dimensions suggesting a hypercomplex space-time. We chose to allocate 4 real dimensions to obtain the 4-dimension space-time implied by eq. 25-B.1. The remaining 28 real dimensions we allocated to additional symmetry groups—namely seven U(2) groups. We call these groups Connection groups since their role is to "connect" fermions residing in different fermion spectrum blocks.

The structure of the seven additional U(2) groups is not specified. We chose to use a reasonable physical principle to allocate them. We believe their role is to "connect" fermions in different blocks since the fermions within each fermion block have "known" interactions. Note that there are initially eight blocks, each with their own set of symmetry groups and corresponding interactions, and initially no interactions between the eight blocks. If there are no block interactions (except gravity), then the Physics of the fermion set is conceptually disjoint. Thus we choose to implement inter-block interactions as in Fig. 25-B.2 following the principle:

A fermion in any block has interactions either directly, or indirectly, with every other fermion in every other block.

Fig. 25-B.2 shows an implementation of this principle. The horizontal lines in Fig. 25-B.2 indicate 1:1 transformations between all corresponding fermions of each "Normal" and "Dark" block. The three "angled" lines indicate 1:1 transformations between corresponding fermions of a "Normal" and a "Dark" fermion block in the layer above it. The result is a see-saw type of pattern.

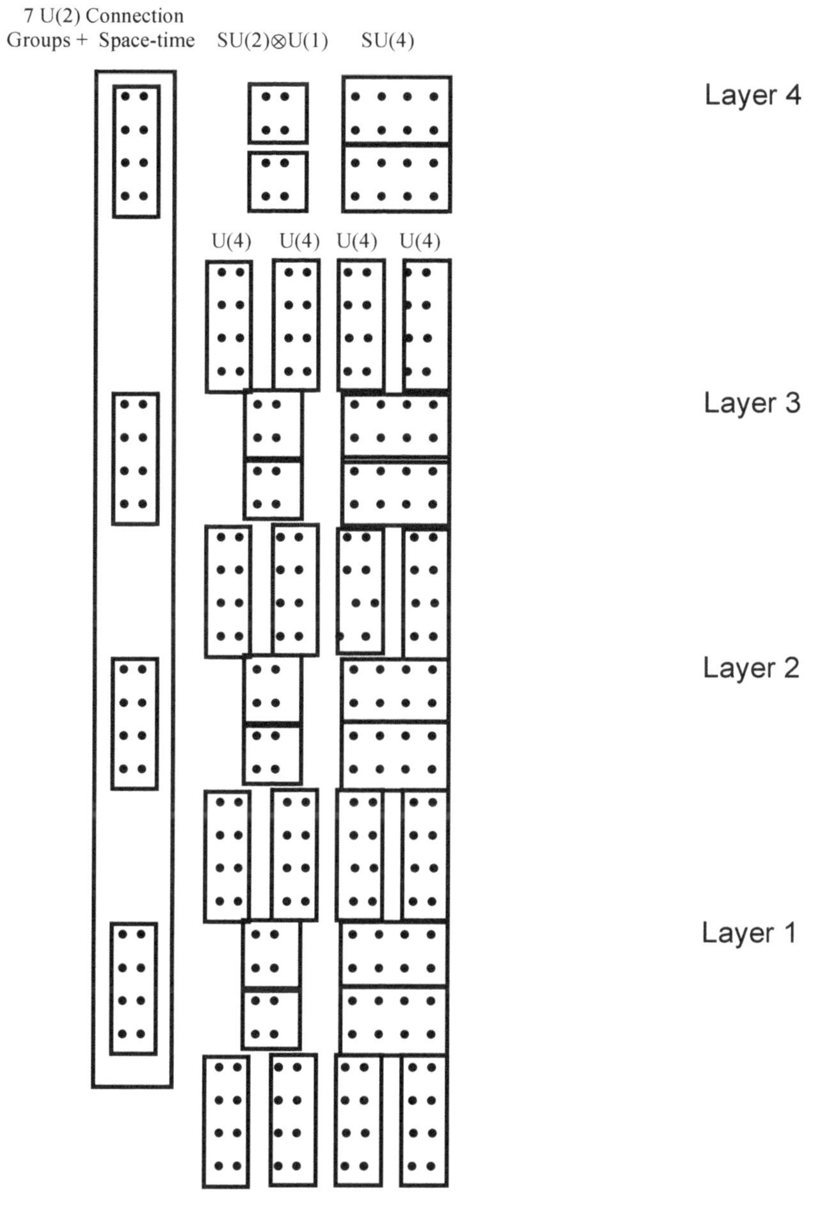

Figure 25-B.1. The four layers of QUeST, UST and QUeST internal symmetry groups (and space-time) with SU(4) before breakdown to SU(3)⊗U(1). Note the left column of blocks are combined below to specify a 4 dimension real space-time plus seven U(2) Connection groups. Note each layer has 64 dimensions = 56 + 8 dimensions.

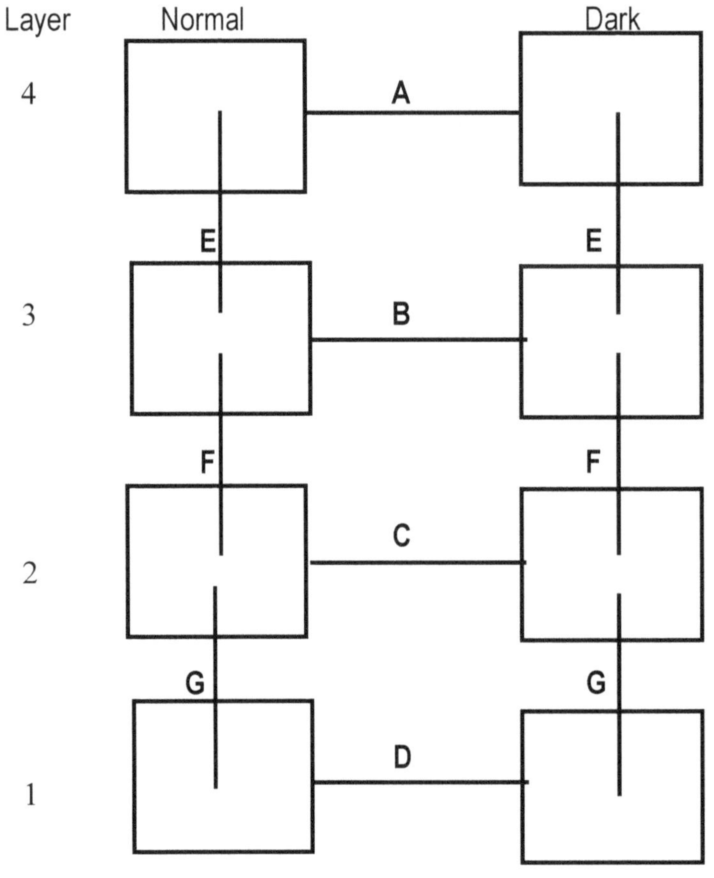

Figure 25-B.2. The seven U(2) Connection groups (shown as 10 lines) between the eight QUeST/UST blocks. Connection groups are obtained by transfering 28 dimensions from QUeST space-time to internal symmetries with the consequent reduction of the space-time from four octonion (complex quaternion) coordinates to four real coordinates. The Connection groups generate rotations and interactions between corresponding fermions and vector bosons of each pair of blocks. The Normal and Dark sector U(2) vertical connections above (E, F, G) represent the same U(2) groups.

25-B.1.1 The U(2) Connection Groups

The seven U(2) Connection groups of Fig. 25-B.2 generate "rotations" and interactions between *corresponding* fermions and vector bosons of each pair of blocks of the eight blocks of fermions in QUeST/UST.

25-B.1.1.1 Horizontal Lines

The horizontal lines in Fig. 25-B.2 (A, B, C, and D) each represent a U(2) Connection group that “rotates” two *corresponding* fermions in the Normal and Dark sectors of each layer. Thus a Normal e is “rotated” with a corresponding Dark e, and so on.

Each of the four horizontal Connection Groups has a reducible U(2) representation D that is the sum[82] of 4*8 = 32 irreducible U(2) representations. We may view the irreducible representations D_j as strung along the diagonal.

$$D = \sum_{j=1}^{32} D_j \tag{25-B.2}$$

for each of the U(2) groups of the four horizontal lines in Fig. 25-B.2.

The U(2) group also specifies gauge field interactions between corresponding fermions in each layer of the Normal and Dark sectors of the form

$$g\overline{\Psi}_{Nn}\gamma\cdot A\cdot T\Psi_{Dn} \tag{25-B.3}$$

where N indicates a Normal fermion and D indicates the corresponding Dark fermion, with A being a U(2) gauge vector boson, and n the label for corresponding fermions.

These U(2) transformations imply that the Normal and Dark sectors have the same species.

25-B.1.1.2 Vertical Lines

The pairs of vertical lines in Fig. 25-B.2 (E, F, G) each represent a U(2) Connection group that “rotates” sets of two *corresponding* fermions in adjacent layers as shown in Fig. 25-B.2 in the Normal and Dark sectors. Thus a Normal e in layer 1 is rotated with a corresponding Normal e in layer 2, and so on.

Each of the three (six counting both Normal and Dark lines in Fig. 25-B.2) vertical Connection Groups has a reducible U(2) representation D that is the sum of 64 irreducible U(2) representations.[83] We may view D as an array of 64 U(2) irreducible representation dimensions D_j strung along the diagonal.

$$D = \sum_{j=1}^{64} D_j \tag{25-B.4}$$

for each of the U(2) groups of the 3 (6) horizontal lines in Fig. 25-B.2. Note the 64 irreducible representations include both Normal and Dark sectors of a layer. [84]

Each U(2) group also specifies a gauge field interaction between corresponding fermions in adjacent layers for both Normal and Dark sectors:

---

[82] Eight fermions per generation ▪ four generations, thus accounting for each fermion in a block.

[83] Eight fermions per generations ▪ four generations ▪ 2 types of matter (Normal and Dark).

[84] There are 64 fermions in total for each of the four layers of NEWQUeST/NEWEST.

$$g\overline{\Psi}_{nl_1}\gamma\cdot A\cdot T\Psi_{nl_2} \tag{25-B.5}$$

where l_1 and l_2 designate layers, A is a gauge field vector boson, and n the label for corresponding fermions.

Each E, F, and G U(2) group reducible representation includes both Normal and Dark sectors.

25-B.1.2 The Connection Groups are UltraWeak Interactions

Since there is no convincing experimental evidence for particle interactions between Normal and Dark matter, or between Normal fermion layers the Connection groups appear to be UltraWeak.

25-B.2 UTMOST with Six Real Space-Time Coordinates (Dimensions)

The UTMOST space is a Blaha number N = 6 HyperCosmos space. It is the Megaverse (Multiverse) with 6 real-valued space-time coordinates by eq.25-B.1 above. It has a $d_{d6} = 1024$ dimension array.

We view the UTMOST dimension array as composed of four copies of the N = 7 dimension array. Initially $4\cdot 32 = 128$ dimensions are for space-time coordinates.

We allocate $4\cdot 28 = 112$ dimensions to each of the four QUeST copies within UTMOST to give each copy 7 SU(2) Connection groups. Therefore 4* 28 = 112 of the 128 space-time dimensions of UTMOST are mapped. The remaining 16 dimensions give 6 real-valued space-time dimensions to the Megaverse space, 8 dimensions to a Megaverse SU(4) Connection group, and 2 dimensions to a U(1) group for all fermions in the Megaverse. See Figs. 25-B.3 and 25-B.4.

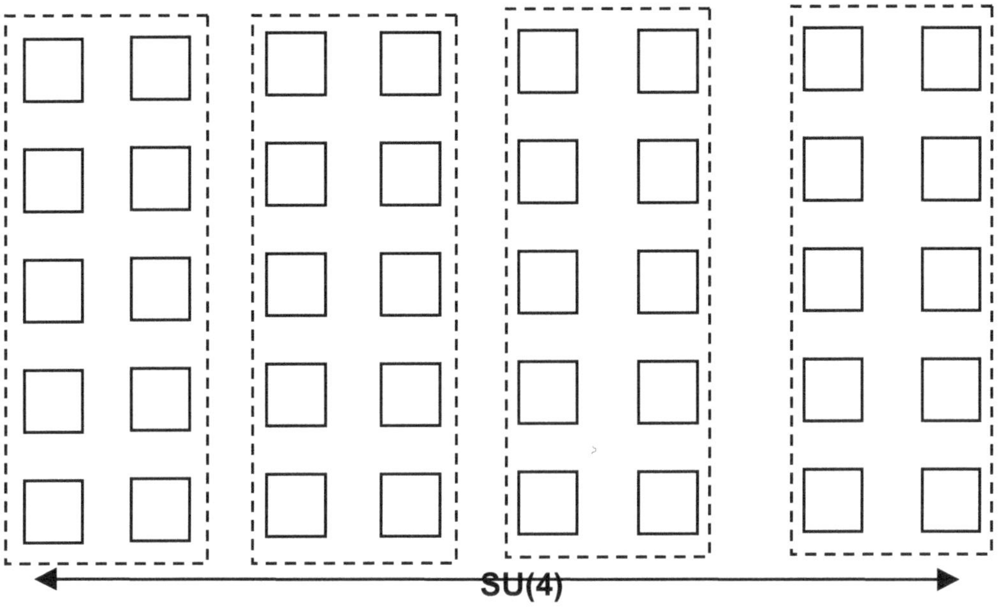

Figure 25-B.3. UTMOST has four QUeST copies. An SU(4) internal symmetry Connection group maps between corresponding fermions in the four copies: fermion by fermion. An additional U(1) Connection group applies to every corresponding fermion. It is not shown in this figure.

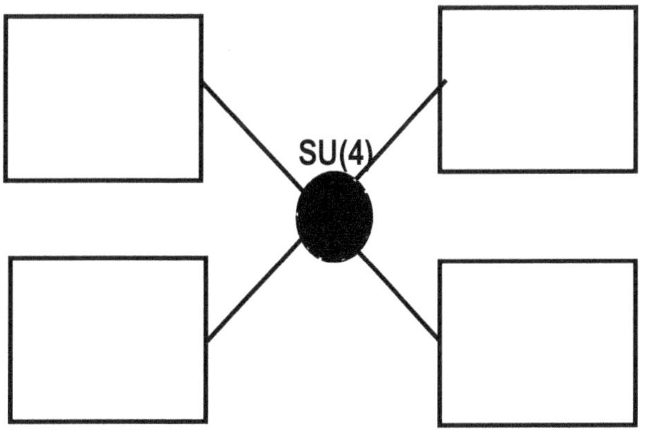

Figure 25-B.4. The SU(4) Connection Group of UTMOST connecting fermions in the four QUeST "copies" blocks. An additional U(1) Connection group applies to every corresponding fermion. It is not shown in this figure.

25-B.3 Maxiverse with Eight Real Coordinates (Dimensions)

The Maxiverse is a Blaha number N = 5 space with a dimension array of d_{d5} = 4096 dimensions.

The Maxiverse contains four copies of UTMOST. Each UTMOST has 4*6 = 24 space-time dimensions. (The remainder in each UTMOST copy consists of internal symmetries and Connection groups.) We allocate the 24 dimensions to eight real space-time dimensions plus 16 dimensions for a new ultraweak (possibly broken) SU(8) Connection group for the four parts of the Maxiverse. See Fig. 25-B.5.

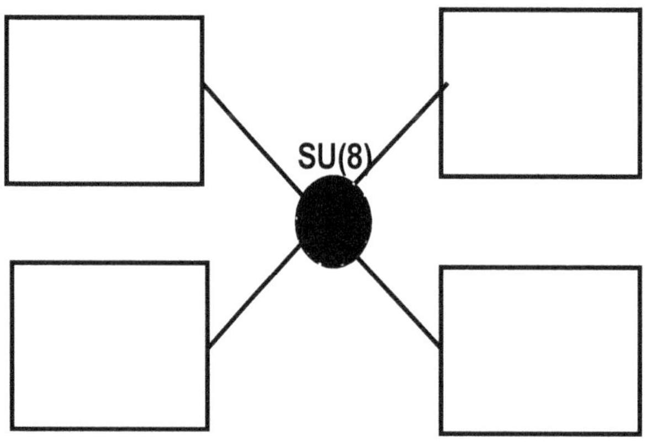

Figure 25-B.5. The SU(8) Connection Group of Maxiverse connecting fermions in the four UTMOST "copies" blocks.

25-B.4 Determining the Connection Groups for a Space

In the previous sections we determined the Connection Groups for N = 7, N = 6, and N = 5 spaces. The dimensions required for the Connection Groups of a space d_{gN} for N ≤ 6 can be obtained by noting the space-time dimension of a space-time r is related to the space-time dimension r – 2 of the next lower space. Since a space is four copies of the next lower space it has 4(r – 2) space-time dimensions from the four duplicates. Then we find

$$4r_{N+1} = 4(r_N - 2) = r_N + d_{gN}$$

or

$$d_{gN} = 3r_N - 8 \qquad (25\text{-B.6})$$

where d_{gN} is the *real-valued* space-time dimensions reallocated to Connection Groups in space N. For N = 6, $r_6 = 6$, $d_{gN} = 10$. For N = 5, $r_5 = 8$, $d_{gN} = 16$. And so on.

Eq. 25-B.6 determines the dimensions available for Connection Groups for the space with Blaha number N ≤ 6. Using eq. 25-B.1:

$$r_N = \log_2 (d_{dN}/16) \qquad (25\text{-B.7})$$

we relate d_{gN} to the size of the dimension array:

$$d_{gN} = 3 \log_2(d_{dN}/16) - 8 \qquad (25\text{-B.8})$$

thus specifying the number of space-time dimensions allocated to space number N.

REFERENCES

Akhiezer, N. I., Frink, A. H. (tr), 1962, *The Calculus of Variations* (Blaisdell Publishing, New York, 1962).

Bjorken, J. D., Drell, S. D., 1964, *Relativistic Quantum Mechanics* (McGraw-Hill, New York, 1965).

Bjorken, J. D., Drell, S. D., 1965, *Relativistic Quantum Fields* (McGraw-Hill, New York, 1965).

Blaha, S., 1995, *C++ for Professional Programming* (International Thomson Publishing, Boston, 1995).

_______, 1998, *Cosmos and Consciousness* (Pingree-Hill Publishing, Auburn, NH, 1998 and 2002).

_______, 2002, *A Finite Unified Quantum Field Theory of the Elementary Particle Standard Model and Quantum Gravity Based on New Quantum Dimensions™ & a New Paradigm in the Calculus of Variations* (Pingree-Hill Publishing, Auburn, NH, 2002).

_______, 2004, *Quantum Big Bang Cosmology: Complex Space-time General Relativity, Quantum Coordinates™ Dodecahedral Universe, Inflation, and New Spin 0, ½, 1 & 2 Tachyons & Imagyons* (Pingree-Hill Publishing, Auburn, NH, 2004).

______, *2005a, Quantum Theory of the Third Kind: A New Type of Divergence-free Quantum Field Theory Supporting a Unified Standard Model of Elementary Particles and Quantum Gravity based on a New Method in the Calculus of Variations* (Pingree-Hill Publishing, Auburn, NH, 2005).

______, 2005b, *The Metatheory of Physics Theories, and the Theory of Everything as a Quantum Computer Language* (Pingree-Hill Publishing, Auburn, NH, 2005).

______, 2005c, *The Equivalence of Elementary Particle Theories and Computer Languages: Quantum Computers, Turing Machines, Standard Model, Superstring Theory, and a Proof that Gödel's Theorem Implies Nature Must Be Quantum* (Pingree-Hill Publishing, Auburn, NH, 2005).

______, 2006a, *The Foundation of the Forces of Nature* (Pingree-Hill Publishing, Auburn, NH, 2006).

______, 2006b, *A Derivation of ElectroWeak Theory based on an Extension of Special Relativity; Black Hole Tachyons; & Tachyons of Any Spin.* (Pingree-Hill Publishing, Auburn, NH, 2006).

_______, 2007a, *Physics Beyond the Light Barrier: The Source of Parity Violation, Tachyons, and A Derivation of Standard Model Features* (Pingree-Hill Publishing, Auburn, NH, 2007).

______, 2007b, *The Origin of the Standard Model: The Genesis of Four Quark and Lepton Species, Parity Violation, the ElectroWeak Sector, Color SU(3), Three Visible Generations of Fermions, and One Generation of Dark Matter with Dark Energy* (Pingree-Hill Publishing, Auburn, NH, 2007).

______, *2008a, A Direct Derivation of the Form of the Standard Model From GL(16) (Pingree-Hill Publishing, Auburn, NH, 2008).*

______, 2008b, *A Complete Derivation of the Form of the Standard Model With a New Method to Generate Particle Masses Second Edition* (Pingree-Hill Publishing, Auburn, NH, 2008)

______, 2009, *The Algebra of Thought & Reality: The Mathematical Basis for Plato's Theory of Ideas, and Reality Extended to Include A Priori Observers and Space-Time Second Edition* (Pingree-Hill Publishing, Auburn, NH, 2009).

______, 2010a, *Operator Metaphysics: A New Metaphysics Based on a New Operator Logic and a New Quantum Operator Logic that Lead to a Mathematical Basis for Plato's Theory of Ideas and Reality* (Pingree-Hill Publishing, Auburn, NH, 2010).

REFERENCES

______, 2010b, *The Standard Model's Form Derived from Operator Logic, Superluminal Transformations and GL(16)* (Pingree-Hill Publishing, Auburn, NH, 2010).

______, 2010c, *SuperCivilizations: Civilizations as Superorganisms* (McMann-Fisher Publishing, Auburn, NH, 2010).

______, 2011a, *21st Century Natural Philosophy Of Ultimate Physical Reality* (McMann-Fisher Publishing, Auburn, NH, 2011).

______, 2011b, *All the Universe! Faster Than Light Tachyon Quark Starships & Particle Accelerators with the LHC as a Prototype Starship Drive Scientific Edition* (Pingree-Hill Publishing, Auburn, NH, 2011).

______, 2011c, *From Asynchronous Logic to The Standard Model to Superflight to the Stars* (Blaha Research, Auburn, NH, 2011).

______, 2012a, *From Asynchronous Logic to The Standard Model to Superflight to the Stars volume 2: Superluminal CP and CPT, U(4) Complex General Relativity and The Standard Model, Complex Vierbein General Relativity, Kinetic Theory, Thermodynamics* (Blaha Research, Auburn, NH, 2012).

______, 2012b, *Standard Model Symmetries, And Four And Sixteen Dimension Complex Relativity; The Origin Of Higgs Mass Terms* (Blaha Reasearch, Auburn, NH, 2012).

______, 2013a, *Multi-Stage Space Guns, Micro-Pulse Nuclear Rockets, and Faster-Than-Light Quark-Gluon Ion Drive Starships* (Blaha Research, Auburn, NH, 2013).

______, 2013b, *The Bridge to Dark Matter; A New Sibling Universe; Dark Energy; Inflatons; Quantum Big Bang; Superluminal Physics; An Extended Standard Model Based on Geometry* (Blaha Reasearch, Auburn, NH, 2013).

______, 2014a, *Universes and Megaverses: From a New Standard Model to a Physical Megaverse; The Big Bang; Our Sibling Universe's Wormhole; Origin of the Cosmological Constant, Spatial Asymmetry of the Universe, and its Web of Galaxies; A Baryonic Field between Universes and Particles; Megaverse Extended Wheeler-DeWitt Equation* (Blaha Reasearch, Auburn, NH, 2014).

______, 2014b, *All the Megaverse! Starships Exploring the Endless Universes of the Cosmos Using the Baryonic Force* (Blaha Research, Auburn, NH, 2014).

______, 2014c, *All the Megaverse! II Between Megaverse Universes: Quantum Entanglement Explained by the Megaverse Coherent Baryonic Radiation Devices – PHASERs Neutron Star Megaverse Slingshot Dynamics Spiritual and UFO Events, and the Megaverse Microscopic Entry into the Megaverse* (Blaha Research, Auburn, NH, 2014).

______, 2015a, *PHYSICS IS LOGIC PAINTED ON THE VOID: Origin of Bare Masses and The Standard Model in Logic, U(4) Origin of the Generations, Normal and Dark Baryonic Forces, Dark Matter, Dark Energy, The Big Bang, Complex General Relativity, A Megaverse of Universe Particles* (Blaha Research, Auburn, NH, 2015).

______, 2015b, *PHYSICS IS LOGIC Part II: The Theory of Everything, The Megaverse Theory of Everything, U(4)⊗U(4) Grand Unified Theory (GUT), Inertial Mass = Gravitational Mass, Unified Extended Standard Model and a New Complex General Relativity with Higgs Particles, Generation Group Higgs Particles* (Blaha Research, Auburn, NH, 2015).

______, 2015c, *The Origin of Higgs ("God") Particles and the Higgs Mechanism: Physics is Logic III, Beyond Higgs – A Revamped Theory With a Local Arrow of Time, The Theory of Everything Enhanced, Why Inertial Frames are Special, Universes of the Mind* (Blaha Research, Auburn, NH, 2015).

______, 2015d, *The Origin of the Eight Coupling Constants of The Theory of Everything: U(8) Grand Unified Theory of Everything (GUTE), S^8 Coupling Constant Symmetry, Space-Time Dependent Coupling Constants, Big Bang Vacuum Coupling Constants, Physics is Logic IV* (Blaha Research, Auburn, NH, 2015).

______, 2016a, *New Types of Dark Matter, Big Bang Equipartition, and A New U(4) Symmetry in the Theory of Everything: Equipartition Principle for Fermions, Matter is 83.33% Dark, Penetrating the Veil of the Big Bang, Explicit QFT Quark Confinement and Charmonium, Physics is Logic V* (Blaha Research, Auburn, NH, 2016).

______, 2016b, *The Periodic Table of the 192 Quarks and Leptons in The Theory of Everything: The U(4) Layer Group, Physics is Logic VI* (Blaha Research, Auburn, NH, 2016).

______, 2016c, *New Boson Quantum Field Theory, Dark Matter Dynamics, Dark Matter Fermion Layer Mixing, Genesis of Higgs Particles, New Layer Higgs Masses, Higgs Coupling Constants, Non-Abelian Higgs Gauge Fields, Physics is Logic VII* (Blaha Research, Auburn, NH, 2016).

______, 2016d, *Unification of the Strong Interactions and Gravitation: Quark Confinement Linked to Modified Short-Distance Gravity; Physics is Logic VIII* (Blaha Research, Auburn, NH, 2016).

______, 2016e, *MoND: Unification of the Strong Interactions and Gravitation II, Quark Confinement Linked to Large-Scale Gravity, Physics is Logic IX* (Blaha Research, Auburn, NH, 2016).

______, 2016f, *CQ Mechanics: A Unification of Quantum & Classical Mechanics, Quantum/Semi-Classical Entanglement, Quantum/Classical Path Integrals, Quantum/Classical Chaos* (Blaha Research, Auburn, NH, 2016).

______, 2016g, *GEMS Unified Gravity, ElectroMagnetic and Strong Interactions: Manifest Quark Confinement, A Solution for the Proton Spin Puzzle, Modified Gravity on the Galactic Scale* (Pingree Hill Publishing, Auburn, NH, 2016).

______, 2016h, *Unification of the Seven Boson Interactions based on the Riemann-Christoffel Curvature Tensor* (Pingree Hill Publishing, Auburn, NH, 2016).

______, 2017a, *Unification of the Eleven Boson Interactions based on 'Rotations of Interactions'* (Pingree Hill Publishing, Auburn, NH, 2017).

______, 2017b, *The Origin of Fermions and Bosons, and Their Unification* (Pingree Hill Publishing, Auburn, NH, 2017).

______, 2017c, *Megaverse: The Universe of Universes* (Pingree Hill Publishing, Auburn, NH, 2017).

______, 2017d, *SuperSymmetry and the Unified SuperStandard Model* (Pingree Hill Publishing, Auburn, NH, 2017).

______, 2017e, *From Qubits to the Unified SuperStandard Model with Embedded SuperStrings: A Derivation* (Pingree Hill Publishing, Auburn, NH, 2017).

______, 2017f, *The Unified SuperStandard Model in Our Universe and the Megaverse: Quarks, ...* , (Pingree Hill Publishing, Auburn, NH, 2017).

______, 2018a, *The Unified SuperStandard Model and the Megaverse SECOND EDITION A Deeper Theory based on a New Particle Functional Space that Explicates Quantum Entanglement Spookiness (Volume 1)* (Pingree Hill Publishing, Auburn, NH, 2018).

______, 2018b, *Cosmos Creation: The Unified SuperStandard Model, Volume 2, SECOND EDITION* (Pingree Hill Publishing, Auburn, NH, 2018).

______, 2018c, *God Theory (*Pingree Hill Publishing, Auburn, NH, 2018).

______, 2018d, *Immortal Eye: God Theory: Second Edition* (Pingree Hill Publishing, Auburn, NH, 2018).

REFERENCES

______, 2018e, *Unification of God Theory and Unified SuperStandard Model THIRD EDITION* (Pingree Hill Publishing, Auburn, NH, 2018).

______, 2019a, *Calculation of: QED α = 1/137, and Other Coupling Constants of the Unified SuperStandard Theory* (Pingree Hill Publishing, Auburn, NH, 2019).

______, 2019b, *Coupling Constants of the Unified SuperStandard Theory SECOND EDITION* (Pingree Hill Publishing, Auburn, NH, 2019).

______, 2019c, *New Hybrid Quantum Big_Bang–Megaverse_Driven Universe with a Finite Big Bang and an Increasing Hubble Constant* (Pingree Hill Publishing, Auburn, NH, 2019).

______, 2019d, *The Universe, The Electron and The Vacuum* (Pingree Hill Publishing, Auburn, NH, 2019).

______, 2019e, *Quantum Big Bang – Quantum Vacuum Universes (Particles)* (Pingree Hill Publishing, Auburn, NH, 2019).

______, 2019f, *The Exact QED Calculation of the Fine Structure Constant Implies ALL 4D Universes have the Same Physics/Life Prospects* (Pingree Hill Publishing, Auburn, NH, 2019).

______, 2019g, *Unified SuperStandard Theory and the SuperUniverse Model: The Foundation of Science* (Pingree Hill Publishing, Auburn, NH, 2019).

______, 2020a, *Quaternion Unified SuperStandard Theory (The QUeST) and Megaverse Octonion SuperStandard Theory (MOST)* (Pingree Hill Publishing, Auburn, NH, 2020).

______, 2020b, *United Universes Quaternion Universe - Octonion Megaverse* (Pingree Hill Publishing, Auburn, NH, 2020).

______, 2020c, *Unified SuperStandard Theories for Quaternion Universes & The Octonion Megaverse* (Pingree Hill Publishing, Auburn, NH, 2020).

______, 2020d, *The Essence of Eternity: Quaternion & Octonion SuperStandard Theories* (Pingree Hill Publishing, Auburn, NH, 2020).

______, 2020e, *The Essence of Eternity II* (Pingree Hill Publishing, Auburn, NH, 2020).

______, 2020f, *A Very Conscious Universe* (Pingree Hill Publishing, Auburn, NH, 2020).

______, 2020g, *Hypercomplex Universe* (Pingree Hill Publishing, Auburn, NH, 2020).

______, 2020h, *Beneath the Quaternion Universe* (Pingree Hill Publishing, Auburn, NH, 2020).

______, 2020i, *Why is the Universe Real? From Quaternion & Octonion to Real Coordinates* (Pingree Hill Publishing, Auburn, NH, 2020).

______, 2020j, *The Origin of Universes: of Quaternion Unified SuperStandard Theory (QUeST); and of the Octonion Megaverse (UTMOST)* (Pingree Hill Publishing, Auburn, NH, 2020).

______, 2020k, *The Seven Spaces of Creation: Octonion Cosmology* (Pingree Hill Publishing, Auburn, NH, 2020).
______, 2020l, *From Octonion Cosmology to the Unified SuperStandard Theory of Particles* (Pingree Hill Publishing, Auburn, NH, 2020).

______, 2021a, *Pioneering the Cosmos* (Pingree Hill Publishing, Auburn, NH, 2021).

______, 2021b, *Pioneering the Cosmos II* (Pingree Hill Publishing, Auburn, NH, 2021).

______, 2021c, *Beyond Octonion Cosmology* (Pingree Hill Publishing, Auburn, NH, 2021).

______, 2021d, *Universes are Particles* (Pingree Hill Publishing, Auburn, NH, 2021).

______, 2021e, *Octonion-like dna-based life, Universe expansion is decay, Emerging New Physics* (Pingree Hill Publishing, Auburn, NH, 2021).

______, 2021f, *The Science of Creation New Quantum Field Theory of Spaces* (Pingree Hill Publishing, Auburn, NH, 2021).

______, 2021g, *Quantum Space Theory With Application to Octonion Cosmology & Possibly To Fermionic Condensed Matter* (Pingree Hill Publishing, Auburn, NH, 2021).

______, 2021h, *21st Century Natural Philosophy of Octonion Cosmology , and Predestination, Fate, and Free Will* (Pingree Hill Publishing, Auburn, NH, 2021).

______, 2021i, *Beyond Octonion Cosmology II : Origin of the Quantum; A New Generalized Field Theory (GiFT); A Proof of the Spectrum of Universes; Atoms in Higher Universes* (Pingree Hill Publishing, Auburn, NH, 2021).

______, 2021j, *Integration of General Relativity and Quantum Theory: Octonion Cosmology, GiFT, Creation/Annihilation Spaces CASe, Reduction of Spaces to a Few Fermions and Symmetries in Fundamental Frames* (Pingree Hill Publishing, Auburn, NH, 2021).

______, 2022a, *New View of Octonion Cosmology Based on the Unification of General Relativity and Quantum Theory* (Pingree Hill Publishing, Auburn, NH, 2022).

______, 2022b, *The Dust Beneath Hypercomplex Cosmology* (Pingree Hill Publishing, Auburn, NH, 2022).

______, 2022c, *Passing Through Nature to Eternity: ProtoCosmos, HyperCosmos, Unified SuperStandard Theory* (Pingree Hill Publishing, Auburn, NH, 2022).

______, 2022d, *HyperCosmos Fractionation and Fundamental Reference Frame Based Unification: Particle Inner Space Basis of Parton and Dual Resonance Models* (Pingree Hill Publishing, Auburn, NH, 2022).

______, 2022e, *A New UniDimension ProtoCosmos and SuperString F-Theory Relation to the HyperCosmos* (Pingree Hill Publishing, Auburn, NH, 2022).

______, 2022f, *The Cosmic Panorama: ProtoCosmos, HyperCosmos,Unified SuperStandard Theory (UST) Derivation* (Pingree Hill Publishing, Auburn, NH, 2022).

_____, 2022g, *Ultimate Origin: ProtoCosmos and HyperCosmos* (Pingree Hill Publishing, Auburn, NH, 2022).

______, 2023a, *UltraUnification and the Generation of the Cosmos* (Pingree Hill Publishing, Auburn, NH, 2023).

______, 2023b, *God and and Cosmos Theory* (Pingree Hill Publishing, Auburn, NH, 2023).

______, 2023c, *A New Completely Geometric SU(8) Cosmos Theory; New PseudoFermion Fields; Fibonacci-like Dimension Arrays; Ramsey Number Approximation* (Pingree Hill Publishing, Auburn, NH, 2023).

______, 2023d, *Newton's Apple is Now the Fermion* (Pingree Hill Publishing, Auburn, NH, 2023).

______, 2023e,*Cosmos Theory: The Sub-Particle Gambol Model* (Pingree Hill Publishing, Auburn, NH, 2023).

_____, 2024a, *Cosmos-Universe-Particle-Gambol Theory* (Pingree Hill Publishing, Auburn, NH, 2024).

_____, 2024b, *Fractal Cosmos Theory* (Pingree Hill Publishing, Auburn, NH, 2024).

_____, 2024c, *Fractal Cosmic Curve: Tensor-Based CosmosTheory* (Pingree Hill Publishing, Auburn, NH, 2024).

_____, 2024d, *The Eternal Form of Cosmos Theory* (Pingree Hill Publishing, Auburn, NH, 2024).

_____, 2024e, *The Eternal Form of Cosmos Theory Third Edition* (Pingree Hill Publishing, Auburn, NH, 2024).

_____, 2024f, *Fundamental Constants of Cosmos Theory and The Standard Model* (Pingree Hill Publishing, Auburn, NH, 2024).

Eddington, A. S., 1952, *The Mathematical Theory of Relativity* (Cambridge University Press, Cambridge, U.K., 1952).

Fant, Karl M., 2005, *Logically Determined Design: Clockless System Design With NULL Convention Logic* (John Wiley and Sons, Hoboken, NJ, 2005).

Feinberg, G. and Shapiro, R., 1980, *Life Beyond Earth: The Intelligent Earthlings Guide to Life in the Universe* (William Morrow and Company, New York, 1980).

Gelfand, I. M., Fomin, S. V., Silverman, R. A. (tr), 2000, *Calculus of Variations* (Dover Publications, Mineola, NY, 2000).

Giaquinta, M., Modica, G., Souchek, J., 1998, *Cartesian Coordinates in the Calculus of Variations* Volumes I and II (Springer-Verlag, New York, 1998).

Giaquinta, M., Hildebrandt, S., 1996, *Calculus of Variations* Volumes I and II (Springer-Verlag, New York, 1996).

Gradshteyn, I. S. and Ryzhik, I. M., 1965, *Table of Integrals, Series, and Products* (Academic Press, New York, 1965).

Heitler, W., 1954, *The Quantum Theory of Radiation* (Claendon Press, Oxford, UK, 1954).

Huang, Kerson, 1992, *Quarks, Leptons & Gauge Fields 2nd Edition* (World Scientific Publishing Company, Singapore, 1992).

Jost, J., Li-Jost, X., 1998, *Calculus of Variations* (Cambridge University Press, New York, 1998).

Kaku, Michio, 1993, *Quantum Field Theory*, (Oxford University Press, New York, 1993).

Kirk, G. S. and Raven, J. E., 1962, *The Presocratic Philosophers* (Cambridge University Press, New York, 1962).

Landau, L. D. and Lifshitz, E. M., 1987, *Fluid Mechanics 2nd Edition*, (Pergamon Press, Elmsford, NY, 1987).

Rescher, N., 1967, *The Philosophy of Leibniz* (Prentice-Hall, Englewood Cliffs, NJ, 1967).

Riesz, Frigyes and Sz.-Nagy, Béla, 1990, *Functional Analysis* (Dover Publications, New York, 1990).

Sakurai, J. J., 1964, *Invariance Principles and Elementary Particles* (Princeton University Press, Princeton, NJ, 1964).

Weinberg, S., 1972, *Gravitation and Cosmology* (John Wiley and Sons, New York, 1972).

Weinberg, S., 1995, *The Quantum Theory of Fields Volume I* (Cambridge University Press, New York, 1995).

INDEX

About the Author

Stephen Blaha is a well-known Physicist and Man of Letters with interests in Science, Society and civilization, the Arts, and Technology. He had an Alfred P. Sloan Foundation scholarship in college. He received his Ph.D. in Physics from Rockefeller University. He has served on the faculties of several major universities. He was also a Member of the Technical Staff at Bell Laboratories, a manager at the Boston Globe Newspaper, a Director at Wang Laboratories, and President of Blaha Software Inc. and of Janus Associates Inc. (NH).

Among other achievements he was a co-discoverer of the "r potential" for heavy quark binding developing the first (and still the only demonstrable) non-Aeolian gauge theory with an "r" potential; first suggested the existence of topological structures in superfluid He-3; first proposed Yang-Mills theories would appear in condensed matter phenomena with non-scalar order parameters; first developed a grammar-based formalism for quantum computers and applied it to elementary particle theories; first developed a new form of quantum field theory without divergences (thus solving a major 60 year old problem that enabled a unified theory of the Standard Model and Quantum Gravity without divergences to be developed); first developed a formulation of complex General Relativity based on analytic continuation from real space-time; first developed a generalized non-homogeneous Robertson-Walker metric that enabled a quantum theory of the Big Bang to be developed without singularities at t = 0; first generalized Cauchy's theorem and Gauss' theorem to complex, curved multi-dimensional spaces; received Honorable Mention in the Gravity Research Foundation Essay Competition in 1978; first developed a physically acceptable theory of faster-than-light particles; first derived a composition of extremums method in the Calculus of Variations; first quantitatively suggested that inflationary periods in the history of the universe were not needed; first proved Gödel's Theorem implies Nature must be quantum; provided a new alternative to the Higgs Mechanism, and Higgs particles, to generate masses; first showed how to resolve logical paradoxes including Gödel's Undecidability Theorem by developing Operator Logic and Quantum Operator Logic; first developed a quantitative harmonic oscillator-like model of the life cycle, and interactions, of civilizations; first showed how equations describing superorganisms also apply to civilizations. A recent book shows his theory applies successfully to the past 14 years of history and to *new* archaeological data on Andean and Mayan civilizations as well as Early Anatolian and Egyptian civilizations.

He first developed an axiomatic derivation of the form of The Standard Model from geometry – space-time properties – The Unified SuperStandard Model. It unifies all the known forces of Nature. It also has a Dark Matter sector that includes a Dark ElectroWeak sector with Dark doublets and Dark gauge interactions. It uses quantum coordinates to remove infinities that crop up in most interacting quantum field theories and additionally to remove the infinities that appear in the Big Bang and generate inflationary growth of the universe. It shows gravity has a MOND-like form without sacrificing Newton's Laws. It relates the interactions of the MOND-like sector of gravity with the r-potential of Quark Confinement. The axioms of the theory lead to the question of their origin. We suggest in the preceding edition of this book it can be attributed to an entity with God-like properties. We explore these properties in "God Theory" and show they predict that the Cosmos exists forever although individual universes (or incarnations of our universe) "come and go." Several other important results emerge from God Theory such a functionally triune God. The Unified SuperStandard Theory has many other important parts described in the Current Edition of *The Unified SuperStandard Theory* and expanded in subsequent volumes.

Blaha has had a major impact on a succession of elementary particle theories: his Ph.D. thesis (1970), and papers, showed that quantum field theory calculations to all orders in ladder approximations could not give scaling deep inelastic electron-nucleon scattering. He later showed the eigenvalue equation for the fine structure constant α in Johnson-Baker-Willey QED had a zero at α = 1 not 1/137 by solving the Schwinger-Dyson equations to all orders in an approximation that agreed with exact results to 4th order in α thus ending interest in this theory. In 1979 at Prof. Ken Johnson's (MIT) suggestion he calculated the proton-neutron mass difference in the MIT bag model and found the result had the wrong sign reducing interest in the bag model. These results all appear in Physical Review papers. In the 2000's he repeatedly pointed out the shortcomings of SuperString theory and showed that The Standard Model's form could be derived from space-time geometry by an extension of Lorentz transformations to faster than light transformations. This deeper space-time basis greatly increases the possibility that it is part of THE fundamental theory. Recently, Blaha showed that the Weak interactions differed significantly from the Strong, electromagnetic and gravitation interactions in important respects while these interactions had similar features, and suggested that ElectroWeak theory, which is essentially a glued union of the Weak interactions and Electromagnetism, possibly modulo unknown Higgs particle features, be replaced by a unified theory of the other interactions combined with a stand-alone Weak interaction theory. Blaha also showed that, if Charmonium calculations are taken seriously, the Strong interaction coupling constant is only a factor of five larger than the electromagnetic coupling constant, and thus Strong interaction perturbation theory would make sense and yield physically meaningful results.

In graduate school (1965-71) he wrote substantial papers in elementary particles and group theory: The Inelastic E- P Structure Functions in a Gluon Model. Phys. Lett. B40:501-502,1972; Deep-Inelastic E-P Structure Functions In A Ladder Model With Spin 1/2 Nucleons, Phys.Rev. D3:510-523,1971; Continuum Contributions To The Pion Radius, Phys. Rev. 178:2167-2169,1969; Character Analysis of U(N) and SU(N), J. Math. Phys. 10, 2156 (1969); and The Calculation of the Irreducible Characters of the Symmetric Group in Terms of the Compound Characters, (Published as Blaha's Lemma in D. E. Knuth's book: *The Art of Computer Programming Vols. 1 – 4*).

In the early 1980's Blaha was also a pioneer in the development of UNIX for financial, scientific and Internet applications: benchmarked UNIX versions showing that block size was critical for UNIX performance, developing financial modeling software, starting database benchmarking comparison studies, developing Internet-like UNIX networking (1982) and developing a hybrid shell programming technique (1982) that was a precursor to the PERL programming language. He was also the manager of the AT&T ten-year future products development database. His work helped lead to commercial UNIX on computers such as Sun Micros, IBM AIX minis, and Apple computers.

In the 1980's he pioneered the development of PC Desktop Publishing on laser printers and was nominated for three "Awards for Technical Excellence" in 1987 by PC Magazine for PC software products that he designed and developed.

Recently he has developed a theory of Megaverses – actual universes of which our universe is one – with quantum particle-like properties based on the Wheeler-DeWitt equation of Quantum Gravity. He has developed a theory of a baryonic force, which had been conjectured many years ago, and estimated the strength of the force based on discrepancies in measurements of the gravitational constant G. This force, operative in D-dimensional space, can be used to escape from our universe in "uniships" which are the equivalent of the faster-than-light starships proposed in the author's earlier books. Thus travel to other universes, as well as to other stars is possible.

Blaha also considered the complexified Wheeler-DeWitt equation and showed that its limitation to real-valued coordinates and metrics generated a Cosmological Constant in the Einstein equations.

The author has also recently written a series of books on the serious problems of the United States and their solution as well as a book on the decline of Mankind that will follow from current social and genetic trends in Mankind.

In the past twenty years Dr. Blaha has written over 80 books on a wide range of topics. Some recent major works are: *From Asynchronous Logic to The Standard Model to Superflight to the Stars, All the Universe!, SuperCivilizations: Civilizations as Superorganisms, America's Future: an Islamic Surge, ISIS, al Qaeda, World Epidemics, Ukraine, Russia-China Pact, US Leadership Crisis, The Rises and Falls of Man – Destiny – 3000 AD: New Support for a Superorganism MACRO-THEORY of CIVILIZATIONS From CURRENT WORLD TRENDS and NEW Peruvian, Pre-Mayan, Mayan, Anatolian, and Early Egyptian Data, with a Projection to 3000 AD,* and *Mankind in Decline: Genetic Disasters, Human-Animal Hybrids, Overpopulation, Pollution, Global Warming, Food and Water Shortages, Desertification, Poverty, Rising Violence, Genocide, Epidemics, Wars, Leadership Failure.*

He has taught approximately 4,000 students in undergraduate, graduate, and postgraduate corporate education courses primarily in major universities, and large companies and government agencies.

He developed a quantum theory, The Unified SuperStandard Theory (UST), which describes elementary particles in detail without the difficulties of conventional quantum field theory. He found that the internal symmetries of this theory could be exactly derived from an octonion theory called QUeST. He further found that another octonion theory (UTMOST) describes the Megaverse. It can hold QUeST universes such as our own universe. It has an internal symmetry structure which is a superset of the QUeST internal symmetries.

Recently he developed Octonion Cosmology. He replaced it with HyperCosmos theory, which has significantly better features. He developed a fractionalization process for dimensions, particles and symmetry groups. He also described transformation that reduced particles and dimensions to a far more compact form. He also developed a precursor theory ProtoCosmos that leads to the HyperCosmos.

The author showed that space-time and Internal Symmetries can be unified in any of the ten HyperCosmos spaces in their associated HyperUnification spaces. The combined set of HyperUnification spaces enable all HyperCosmos dimensions to be obtained by a General Relativistic transformation from one primordial dimension in the 42 space-time dimension unified HyperUnification space.

At present the author devel;oped the Cosmos Theory that incorporates ProtoCosmos Theory, HyperCosmos Theory, Limos Theory, Second Kind HyperCosmos Theory and HyperUnification Spaces. He has introduced PseudoFermion wave functions and theory, He has related Cosmos Theory to Regge trajectories of spaces, parton theory, Veneziano amplitudes, Fibonacci numbers and Ramsey numbers. He has calculated an approximation to the difficult R(n,n) Ramsey numbers.

He has developed a Gambol Model that successfully accounts for e-p deep inelastic scattering, fundamental particle resonances, hadron scattering, and the inner structure of particles based on confinement through Casimir forces of ideal gambol gases. The Gambol Planckian Distribution was derived.

He has applied the Gambol Model to particles, universes, and the Cosmos of universes. He showed that the Cosmos may have a distribution of 23 universes corresponding to various Cosmos spaces.

Recently he showed that Cosmos Theory follows from the number of independent asymmetric tensors in a dimension r. He also showed the close parallel between the form of γ-matrices and Cosmos Theory dimension arrays. The closeness suggested that dimension arrays have the same importance as γ-matrices for fermions.

He demonstrated that the pressure of fermions within a space of dimension r balances the Casimir vacuum energy force for 18 dimensions. He showed that $2e\pi$ = 17.02 marks the critical point where pressure balances Casimir force, which implies r = 18 is the highest dimension Physical Cosmos space. The dimension $2e\pi$ appears to set the approximate dimension for Cosmos spaces with dimension array size $2^{r+4} \cong (17.02/8)^{r+4}. \cong (e\pi/4)^{r+4}. \cong 2.13^{r+4}$.

Now he has found the sequence of Coupling Constant values in the Standard Model and UST.

www.ingramcontent.com/pod-product-compliance
Ingram Content Group UK Ltd.
Pitfield, Milton Keynes, MK11 3LW, UK
UKHW060119010826
14089UKWH00005B/67

9 798989 408467